Compendium of Hop Diseases and Pests

Edited by

Walter F. Mahaffee
USDA-ARS
Horticulture Crops Research Laboratory, Corvallis, OR

Sarah J. Pethybridge
Botanical Resources Australia
Agricultural Services Pty. Ltd., Ulverstone, Tasmania, Australia

David H. Gent
USDA-ARS
Forage Seed and Cereal Research Unit, Corvallis, OR

APS PRESS

The American Phytopathological Society

Front cover photograph courtesy David H. Gent. Back cover photographs courtesy Stephen Ausmus (lower right) and Walter F. Mahaffee (all others)

Reference in this publication to a trademark, proprietary product, or company name by personnel of the U.S. Department of Agriculture or anyone else is intended for explicit description only and does not imply approval or recommendation to the exclusion of others that may be suitable.

Library of Congress Control Number: 2009923813
International Standard Book Number: 978-0-89054-376-4

Printed in the United States of America on acid-free paper.

The American Phytopathological Society
3340 Pilot Knob Road
St. Paul, Minnesota 55121, U.S.A.

Preface

This compendium of disease, arthropod pests, and disorders of hop (*Humulus lupulus* L.) is designed to be a practical reference for anyone interested in growing hops, whether for commercial production or for ornamental or home use. It is written as a general reference for individuals with limited training in entomology, plant pathology, and horticultural science but is intended to have sufficient detail to be useful for individuals with more advanced training or expertise. Each section was written and reviewed by recognized experts on the various subjects from across the world.

The compendium attempts to be inclusive of all diseases and pests known to occur throughout the world's hop production regions. It builds upon other key references for this crop including *Hops* (R. A. Neve, 1991, published by Chapman & Hall, New York) and *The Hop Atlas: The History and Geography of the Cultivated Plant* (H. J. Barth, C. Klinke, and C. Schmidt, 1994, John Barth & Son, Nuremberg, Germany).

The introduction consists of information on the genus *Humulus*, unique features of commercial and home hop production, harvest, use and chemistry of hop cones, and characteristics of world cultivars. Diseases caused by bacteria, fungi, nematodes, oomycetes, viruses, and phytoplasmas are discussed in Part I. Numerous arthropod pests are discussed in Part II. Part III addresses postharvest, nutrient, and other disorders. A glossary of terms used in plant pathology, mycology, entomology, and general hop production is also included.

The authors of each section are listed at the end of the individual sections. Sources of images and illustrations are indicated in the captions. We thank the individuals who contributed or reviewed sections or the entire manuscript, as well as numerous individuals and organizations within the hop industry who have provided support, knowledge, and images.

Walter F. Mahaffee
Sarah J. Pethybridge
David H. Gent

Contributors and Reviewers

A. M. Alvarez, University of Hawaii, Honolulu, Hawaii
E. Atkins, Bonners Ferry, Idaho
D. J. Barbara, Horticulture Research International, Warwick, United Kingdom
J. D. Barbour, University of Idaho, Parma, Idaho
R. A. Beatson, HortResearch, Motueka, New Zealand
J. C. Bienapfl, University of California, Davis, California
D. Bruck, United States Department of Agriculture, Agricultural Research Service, Corvallis, Oregon
C. T. Bull, United State Department of Agriculture, Agricultural Research Service, Salinas, California
A. J. Dreves, Oregon State University, Corvallis, Oregon
K. C. Eastwell, Washington State University, Prosser, Washington
B. Engelhard, Bavarian State Research Center for Agriculture, Wolnzach, Germany
G. C. Fisher, Oregon State University, Corvallis, Oregon
D. H. Gent, United States Department of Agriculture, Agricultural Research Service, Corvallis, Oregon
T. Grasswitz, Washington State University, Prosser, Washington
Ken Gray Image Collection, Oregon State University
G. G. Grove, Washington State University, Prosser, Washington
M. Grudzińska-Sterno, Institute of Soil Science and Plant Cultivation, Puławy, Poland
F. S. Hay, University of Tasmania, Burnie, Tasmania, Australia
P. Hamilton, Hop Products Australia, Hobart, Tasmania, Australia
J. A. Henning, United States Department of Agriculture, Agricultural Research Service, Corvallis Oregon
D. Hysert, John I. Haas, Inc., Yakima, Washington State
D. G. James, Washington State University, Prosser, Washington
D. A. Johnson, Washington State University, Pullman, Washington
M. Kamińska, Institute of Soil Science and Plant Cultivation, Puławy, Poland
S. T. Kenny, Washington State University, Prosser, Washington
A. Lawrie, The Royal Melbourne Institute of Technology, Melbourne, Victoria, Australia
W. F. Mahaffee, United States Department of Agriculture, Agricultural Research Service, Corvallis, Oregon
P. McGee, Australian Food and Grocery Council, Australian Capital Territory, Australia
T. Neill, United States Department of Agriculture, Agricultural Research Service, Corvallis Oregon
M. E. Nelson, Washington State University, Prosser, Washington
C. M. Ocamb, Oregon State University, Corvallis, Oregon
R. Parker, Washington State University, Prosser, Washington
S. J. Pethybridge, Botanical Resources Australia – Agricultural Services Pty. Ltd., Ulverstone, Tasmania, Australia
I. Porter, Department of Primary Industries, Knoxfield, Victoria, Australia
S. Radišek, Slovenian Institute for Hop and Breeding, Žalec, Slovenia
T. Sano, Hirosaki University, Japan
C. B. Skotland, Washington State University, Prosser, Washington
H. Śliwa, Institute of Soil Science and Plant Cultivation, Puławy, Poland
D. R. Smith, Yakima, Washington
J. T. Smith, HortEye Ltd.. New Zealand
E. Solarska, Institute of Soil Science and Plant Cultivation, Puławy, Poland
J. R. Umble, Fall Creek Farm and Nursery, Inc., Lowell, Oregon
J. Vostrel, Hop Research Institute Co., Ltd., Žatec, Czech Republic
F. Weihrauch, Bavarian State Research Center for Agriculture, Wolnzach, Germany
D. Whitener, G. S. Long Co., Union Gap, Washington
C. R. Wilson, University of Tasmania, New Town, Tasmania, Australia

Contents

Introduction

The Genus *Humulus*

Humulus is a genus of dioecious (rarely monoecious) flowering plants indigenous to the temperate zones of the Northern Hemisphere, with China considered the center of origin. It is one of two genera (*Humulus* and *Cannabis*) that belong to the family Cannabaceae and contains three species: *H. japonicus* Siebold & Zucc., *H. lupulus* L., and *H. yunnanensis* Hu.

All *Humulus* species are short-day, herbaceous, climbing bines (stout stems with stiff and abundant hooked trichomes that assist in climbing instead of tendrils, suckers, or other appendages as with vines). Bines twine in a clockwise direction around structural supports (e.g., trees and bushes). Male and female plants are indistinguishable until flowering.

H. lupulus (common hop) is a perennial that is native to Europe, eastern Asia, and North America. The species is subdivided into five botanical varieties distinguished by morphogeographic characters. Both sexes contain 20 chromosomes and are generally diploid with a recognizable Y chromosome.

H. lupulus var. *neomexicanus* A. Nelson & Cockerell appears to be adapted to the western Cordilleran conditions of North America. Plants possess leaves with relatively more lobes, deeper lobe clefts, and the greatest density of abaxial leaf glands compared with the other varieties.

H. lupulus var. *pubescens* E. Small is found in the American Midwest. The leaves tend to be entire to three-lobed with more pronounced marginal serration than other *Humulus* species, abundant pubescence on the undersides, and numerous climbing hairs on the petioles.

H. lupulus var. *lupulus* is distributed in Europe, Asia, and Africa and has been introduced into eastern North America and other areas. Compared with other varieties, it has the fewest hairs and glands on the undersides of the leaves. The density and height of the climbing hairs on the petioles and the frequency of marginal serrations on the leaves are usually less than that of the other varieties.

H. lupulus. var. *cordifolius* (Miq.) Maxim. is distributed in eastern Asia, mainly in Japan, and tends to have climbing hairs that are taller but less dense than those of other varieties.

H. lupulus var. *lupuloides* E. Small is found in eastern and central North America and encompasses *Humulus* plants of North America that do not key to any of the other varieties.

In geographic regions where the botanical varieties of *H. lupulus* overlap, there is the potential for introgression. This has been reported between *H. lupulus* var. *lupuloides* and *H. lupulus* var. *neomexicanus* near Manitoba, Canada. There is also evidence of introgression among species in China.

Only the female plants of *H. lupulus* are agronomically important for the resins, essential oils, and polyphenols contained within the lupulin glands (glandular trichomes) on cones (strobiles), which are used to flavor and preserve beer. All commercial cultivars of hop are *H. lupulus*, and most were derived from the European-type hop crossed with germplasm from North America.

H. japonicus (syn. *H. scandens* (Lour.) Merr.) is an annual. Leaves are arranged palmately, have five to nine lobes, and are covered in strong, hooked, climbing hairs. Flowering occurs sequentially from June to October (Northern Hemisphere). Cone-like strobiles are similar in appearance to those of *H. lupulus* but have few if any lupulin glands. *H. japonicus* has 16 chromosomes in female plants and 17 chromosomes in male plants. Males also have a recognizable sex chromosome. This species is native to China, Japan, and Taiwan and was introduced into Europe, Russia, and eastern North America as an ornamental plant. It is now recognized as a mildly invasive weed species in the eastern United States.

H. yunnanensis is a perennial with three- to five-lobed palmate leaves, sometimes simple and with dense pubescence on the underside. Flowers and cones are similar to those of *H. lupulus*, except the veins are somewhat raised on the bracts. *H. yunnanensis* is native to high-altitude regions in southern China. No accessions of *H. yunnanensis* reside within the United States Department of Agriculture (USDA) National Plant Germplasm System, and little information exists regarding the botany and ecology of this rare species.

Uses

Throughout history, hop has been used for many purposes. Hops appear to have been recognized as having medicinal properties as early as 3,000 B.C. but were not used in beer until much later. Young hop shoots and leaves were used in salads during medieval times and have recently had a resurgence as so-called "hop asparagus." Cones have been used as pillow stuffing to help with insomnia and in teas and other liquids as remedies for various ailments including nervous stomachs and insomnia. Leaves and cones have been ground and used as a dye for cloth. Hop bines have been used for fiber, decorations, and wreaths. Chemical fractions of hops have also been used as preservatives in sugar processing, ethanol production, cosmetics, and antibiotic replacements in animal feed.

Today, hops are used primarily in flavoring, preserving, and clarifying beer. It is likely that the first use of hops in beer occurred sometime during the fourth to seventh centuries A.D. There is documented evidence that hop was cultivated as a crop in 736 A.D. and that brewers sought and obtained hops in 822 A.D. The use of hops in beer was probably derived from the practice of mixing various medicines in beer before being administered. Further evidence indicates hop was cultivated in 859 A.D. in Bavaria, Germany, although the purpose of this production is unclear.

Today, hops are an essential ingredient in beer. If hops are not used, then by definition the term "beer" cannot be applied and instead the product is termed a malt beverage. In 2005, the world hop production was in excess of 94,000 t from 50,453 ha, and it was used to produce 160 billion liters of beer.

Botanical Characters

Bines of perennial hop plants are capable of growing longer than 7.5 m in a single season from rhizomes. In their natural habitats, hop plants are generally found in riparian areas or other areas that provide sufficient water for growth, structure for climbing, and at latitudes of 25–70°.

The mature root system of hop consists of two morphological types of roots: lateral spreading roots (2–5 m) that are wiry and extensively branched resulting in a fibrous layer in the top 20–30 cm of the soil, and vertical roots (2–3 m in most soils) that are fleshy, irregularly swollen, brittle, and rarely branched that originate from the crown rhizomes. Rhizomes are formed when the base of a seedling or rooted stem swells below the soil line. Over time, the tissue becomes fibrous and spongy with a thick, reddish-brown bark and numerous buds from which shoots will emerge (Fig. 1). The rhizome serves as the overwintering structure for perennial hop plants.

In early spring, a profusion of shoots (long, slender, hexagonal, hollow stems lined with pith tissue) begins to emerge from buds on the rhizomes (Fig. 2). A shoot grows vertically for the first four to seven nodes and then begins to rotate clockwise. The elongating bine grows laterally along the ground until it contacts a structural support and begins to climb by twining around the support. Climbing is enhanced by an abundance of stout, hooked trichomes that are high in silicate and located on the ridges of the hexagonal stem (Fig. 3). The color of the bine varies among cultivars and ranges from reddish purple through degrees of mottling to various shades of green. Lateral stems originate from axil buds at each node of the main stem (Fig. 4).

Leaves are decussate or occasionally tricussate, arising from nodes on the main bine and lateral branches. They are simple with serrated margins, have a cordate base, and are ovate to palmate with numerous lobes (Fig. 5). Petioles and leaf surfaces also have silicate-rich trichomes. The abaxial leaf surface has glandular trichomes containing resins, polyphenols, and essential oils and has varying degrees of pubescence (Fig. 5). Leaf color varies among cultivars and encompasses shades of yellow, purple, and green.

The male inflorescence is a profusely branched panicle that originates from an axil bud of a lateral branch. The individual flowers are small (5–6 mm in diameter). Each has five petals with adherent stamens and anthers and lupulin glands (Fig. 6).

The female inflorescence (commonly called a burr) originates from an axillary bud of a lateral branch and consists of 20–60 flowers on a short central rachis (Figs. 7 and 8) with each node having a pair of bracts. Each bract cup has an ovary with one ovule, which bears two papillated filamentous stigmas without styles. Once fertilized, the stigmas rapidly die while the bracts and bracteoles enlarge and the rachis (strig) elongates and thickens to form strobiles. Occasionally, leaves also develop from the rachis (phyllody) and are referred to as "leafy cones," "angel wings," or "cock hops" (Fig. 9). If unfertilized, the stigmas elongate for about 2–3 weeks. The bract, bracteoles, and rachis enlarge but not to the same extent as do fertilized flowers. Cones consisting of unfertilized flowers are denser or more compact than those that have been fertilized. During this enlargement, the lupulin glands (glandular

Fig. 1. Hop root system. (Courtesy W. F. Mahaffee)

Fig. 2. Hop shoots emerging from dormant crowns in spring. (Courtesy D. H. Gent)

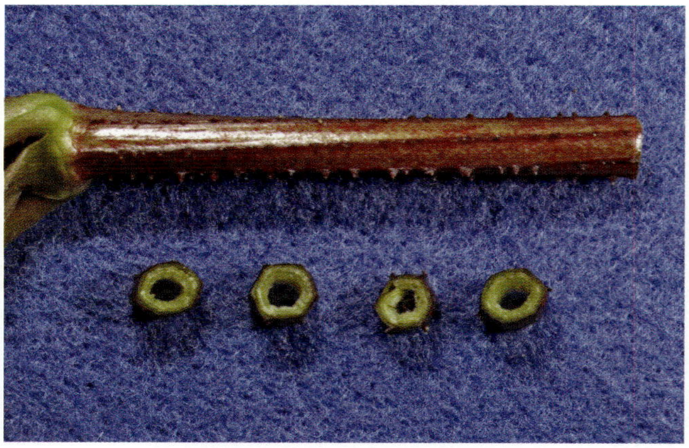

Fig. 3. The twisting hexagonal main bine with the large trichomes that facilitate climbing. (Courtesy W. F. Mahaffee)

Fig. 4. Lateral stem development. (Courtesy D. H. Gent)

Fig. 5. Hop leaves. Left, morphological and color differences; center, adaxial surface; right, abaxial surface. (Courtesy W. F. Mahaffee)

Fig. 6. Male inflorescence of the hop plant. (Courtesy D. H. Gent [top] and W. F. Mahaffee [bottom])

trichomes) develop and begin to accumulate resins and essential oils. Lupulin glands are more abundant on bracteoles than on bracts. The density and composition of the lupulin glands vary greatly among commercial cultivars, production regions, and environmental conditions of the growing season.

Growth and Development

In early spring, shoots begin to emerge from hills (a single plant or multiple plants planted together). The number of shoots is dependent on the size of the crown and root system, the severity of spring pruning, and the cultivar. The bines grow rapidly, as much as 15–25 cm per day, and may reach 5 m or more by the summer solstice. About this time, lateral branches begin to develop in response to the photoperiod. *Humulus* plants respond to decreasing day length and temperature interactions by initiating flowering within weeks of the summer solstice. If the day length at this time is too short (<15 h), flowering is greatly reduced or the plant may not flower at all. This requirement generally relegates hop production to regions above 35° latitude. However, hop can be grown commercially at latitudes less than 35° if supplied with supplemental artificial lighting, as is done in southern South Africa (34° latitude) and Zimbabwe (18° latitude). Male plants tend to flower earlier than female plants.

After flowering, cones develop rapidly, regardless of fertilization. However, fertilized cones are longer and heavier. The lupulin glands accumulate much of their contents within the last few weeks before cones become ripe and ready for harvest. Once ripe, cones may deteriorate rapidly. The bract and bracteoles senesce, and abscission zones form between these tissues and the rachis (strig). The cones then shatter, and the seeds are distributed cupped within a bract.

Germplasm Resources

There is an extensive collection of *Humulus* germplasm resources held by the USDA Agricultural Research Service

Fig. 7. Female inflorescence or burr of the hop plant. (Courtesy D. H. Gent [left and center] and W. F. Mahaffee [right])

Fig. 8. Mature hop cones on side arms (top) and dissected (bottom). In the dissected cone (bottom) from left to right visible are the yellow lupulin glands, the rachis (strig), and bracteoles (lower) and bracts (upper) with seeds. (Courtesy W. F. Mahaffee)

Fig. 9. Abnormal leaf development from the rachis in the female inflorescence (commonly termed "angel wings"). (Courtesy W. F. Mahaffee)

(USDA-ARS) National Clonal Germplasm Repository in Corvallis, Oregon, that consists of commercial cultivars and selections, species material, seed accessions, and tissue-cultured plantlets. The collection is particularly strong in seed accessions of varieties of *H. lupulus* that are found in the United States. All material is freely available upon request (National Clonal Germplasm Repository, USDA-ARS, 33447 Peoria Road, Corvallis, OR 97330).

Hops as a Commodity

Historically, there has been extreme price volatility in response to rising and falling production levels. This volatility is largely because of the limited number of markets and storageability of hops. Until recently, the brewing industry was the only significant market for hops, and whole or pelleted hops had limited storage life. In years when supply exceeded demand, buyers were able to control prices, while in years of shortfalls, prices benefited the growers. This situation created "boom and bust" scenarios that have led to numerous government controls over prices through market orders and other regulations that, for periods of time, stabilized markets and maintained higher prices that allowed some hop growers to prosper. Hops are showing promise for use in sugar processing and as a preservative in ethanol production (currently 3–5% of the world crop). Use as an antimicrobial additive in animal feed could further increase this portion.

The technological advances in storage of processed hop cones and extraction of the resins, polyphenols, and essential oils combined with consumer demand for beers with less bittering traits have reduced the demand for hops. These forces create another unique aspect of hop production: it is very difficult to stop or start growing hops because of the large capital investment needed for very unique and specialized equipment (e.g., harvesting equipment and kilns) and associated infrastructure (U.S.$2–4 million to grow 100 ha). The growth habit of the crop and high labor demands at very specific and short time periods restrict the ability to economize with increasing scale, thus making it difficult to get a significant return on capital investments in a volatile market.

Selected References

Barth, H. J., Klinke, C., and Schmidt, C. 1994. The Hop Atlas: The History and Geography of the Cultivated Plant. Joh. Barth & Sohn, Nuremberg.

Burgess, A. H. 1964. Hops: Botany, Cultivation and Utilization. World Crop Books, Interscience Publication, New York.

Gent, D. H., Nelson, M. E., George, A. E., Grove, G. G., Mahaffee, W. F., Ocamb, C. M., Barbour, J. D., Peetz, A., and Turechek, W. W. 2008. A decade of hop powdery mildew in the Pacific Northwest. Online. Plant Health Progress doi:10.1094/PHP-2008-0314-01-RV.

Murakami, A., Darby, P., Javornik, B., Pais, M. S. S., Seigner, E., Lutz, A., and Svoboda, P. 2006. Molecular phylogeny of wild hops, *Humulus lupulus* L. Heredity 97:66-74.

Neve, R. A. 1991. Hops. Chapman and Hall, London.

Wilson, D. G. 1975. Plant remains from the Graveney boat and the early history of *Humulus lupulus* L. in W. Europe. New Phytol. 75:627-648.

(Prepared by W. F. Mahaffee and S. J. Pethybridge)

Hop Production

Climate and Soils

Cultivated hop plants are produced in a diversity of climates, including cool, maritime regions, humid continental, subtropical, and semiarid environments in both the Northern and Southern Hemispheres. Individual cultivars vary widely in their tolerance of and response to environmental conditions, such as temperature extremes and day length. Consequently, yield and brewing quality of a particular cultivar or cultivar type (aroma or alpha-acid) may not be suitable for production in all climates or areas. Hop culture requires a climate cold enough to satisfy chilling requirements for winter dormancy because crown buds require temperatures below 5°C for at least 5–6 weeks for optimum spring growth. Insufficient chilling may cause plants to break dormancy slowly or erratically, as observed in California and Kenya when hops were grown in those regions. Moreover, crowns can survive temperatures of –25°C or lower when well covered by soil or snow. However, hop plants also require sufficient heat accumulation in spring and summer to favor rapid plant growth and development of aroma and bittering compounds within a relatively short growing season. In general, hop production is favored in climates dominated by dry to moderately wet conditions associated with moderate to warm temperatures during spring and summer.

Hop can be found growing in an array of soil types, ranging from deep alluvial loams, slightly to moderately calcareous eolian silts, and clay-loam soils derived from lacustrine deposits. Commercial production requires deep, well-drained, and friable soils that allow frequent traffic by farm equipment for cultural practices and development of the perennial root system, which can extend to depths of 4 m or more. Soils with a pH near 6.5 are optimal, although the association of surface pH to cone yield and quality is somewhat unclear. Soil amendment is recommended when pH is less than 5.7 or greater than 7.5 to avoid nutrient toxicities or deficiencies, particularly from manganese and zinc.

Propagation

There are several methods by which hop can be propagated. One of the most common methods is propagation by vegetative softwood cuttings, which are approximately 5–8 cm long and consist of one to two nodes and one or two leaves. The cuttings are placed in a peat/sand mixture or floral foam and allowed to root under a mist system or in trays of water. Strap cutting, an alternative propagation method, involves placing ("hilling") soil around and over bines late in the season, which stimulates the development of perennial buds and rhizomatous tissue (Fig. 10). Rhizome pieces with new buds are then removed and planted elsewhere. Propagation is also achieved by layering, in which bines are laid on the ground and covered with soil and the tip is retrained along another string. This allows cuttings to be made between each node once fibrous roots and buds have developed. Many destructive pathogens are readily disseminated in infected propagation materials, and with any propagation method it is important to select planting materials tested and known to be free from pathogens.

Planting and Trellising

Various planting patterns have been used for hop, with plants arranged on a regular grid of 2.1–2.4 m between plants being the most common. Recently, hop yards have been established with plants approximately 1.1–2.5 m apart within rows and 3.6–4.9 m apart between rows to facilitate drip irrigation systems and improve efficiency of cultivation and other cultural practices. Depending on the planting pattern, there are approximately 1,125 to 2,250 plants per hectare, although plant density can be even greater in dwarf hop production. In regions where hops are hand-harvested (e.g., China and Japan), other planting patterns and densities are used. Hop plants have been reported to produce for 100 years or longer, although hop yards are typically replaced every 10 to 20 years or less as a consequence of declining yields or changing market demands by brewers.

In traditional production, hop plants are grown under a trellis system utilizing heavy-gauge wire suspended by poles. The trellis system provides support for the climbing bines, which later produce lateral branches where the cones are borne. Most hops in the United States are grown with a 5.5-m trellis height, although heights of 7 m or taller are common in continental Europe. Trellis height can affect yield, and cultivars with particularly low or high vigor may produce greater yields if grown on a slightly shorter or taller trellis, respectively. Trellis height appears to have been selected on the basis of historical practices and practical considerations for standardizing equipment and field operations. For home growers, trellis height is less critical, since yield is typically of secondary importance, and plants can be grown on a small trellis or on single poles. Hop

Fig. 10. Hop plant hilled mid-season to encourage rhizome development. (Courtesy D. H. Gent)

plants also can be planted closely (~1 m) to form an ornamental hedgerow in landscapes.

In the United Kingdom, and recently in continental Europe and North America, production of traditional or shorter "dwarf" hop cultivars on low trellises (approximately 2–3 m high) is expanding in attempts to reduce production costs, particularly labor (Fig. 11). Yields of traditional cultivars on low trellises are often reduced, but improved production methods may result in improved profit margins thanks to reduced input costs.

Agronomic Practices

In the United States and Europe, bines from the previous season, buds, and young shoots are removed in late winter or early spring by a variety of methods including "pruning" (removal of shoots prior to training), "crowning" (removal of the top 2–5 cm of the crown prior to budbreak), and "scratching" (use of a device that is most similar to a harrow with two spinning disks with tongs that scratch the soil as they spin, usually removing the buds from the crowns within 2–5 cm of the soil surface) (Fig. 12). The timing of spring pruning is largely cultivar specific and can be critical in determining yield potential since it affects the timing of training and thus environmental conditions during vegetative growth and flowering. Spring pruning also helps to reduce inoculum of the downy mildew and powdery mildew pathogens and is essential in regions where these diseases are endemic. Spring pruning can be done mechanically with a tractor-drawn, modified mower deck to cut away the previous season's growth and surface crown buds or with a specialized implement with spinning steel tines to remove the young shoots and bines left from the prior season. With the former method, growers typically "hill-up" soil on top of the crowns near midseason to encourage development of roots and rhizomes near the top of the crown. An additional benefit of hilling soil on crowns is some suppression of downy mildew in the current season because diseased shoots near the crown are buried. Various chemical desiccants (e.g., carfentrazone-ethyl, diquat, and paraquat) can also be used to remove young shoots, with or without a prior mechanical operation to reduce the density of the plant material. For the home gardener, shoots can be removed manually with pruning shears or a garden hoe.

After pruning in early spring, two to four strings of coconut fiber, paper, metal wire, or plastic are tied to the wires on the trellis and anchored to the hills with or without the aid of a small metal clip in a practice referred to as "stringing" (Fig. 13). Stringing is accomplished by manual labor (Fig. 14), although automated stringing machines have been developed. Later in the spring, two to four bines approximately 0.5 m in length are trained onto each string by manually winding bines in a clockwise direction (Fig. 15), although certain cultivars with high vigor may partially self-train. Minimal or no training is needed in dwarf hop production. Selecting the proper training date can be critical for maximizing yield because of cultivar-specific relationships between day length and heat accumulation that influence the time of flowering. Another consideration in selecting training date is disease control, since early training may favor more severe outbreaks of certain diseases. After training, hop bines climb the string and may grow up to 25 cm per day, causing strings to sag under the weight of the developing bines. When plant rows are spaced narrowly, the bines are tied together ("arched") approximately 1.5–2 m above the ground in late spring to allow tractors to drive through the yard for cultural practices and pesticide applications.

As the trained bines grow up the string, superfluous growth of leaves and lower lateral branches is usually removed from the lower 1.5 m of the bine (by a process called "stripping") to minimize the spread of downy and powdery mildew up the canopy. Stripping also increases airflow in the hop yard and

Fig. 11. Dwarf hop plants grown on a 2-m trellis in the United Kingdom. (Courtesy W. F. Mahaffee)

Fig. 12. Hop plant in early spring after mechanical crowning to remove the previous season's shoots and crown buds. (Courtesy D. H. Gent)

Fig. 13. Anchoring strings to a hop hill during stringing operations. (Courtesy D. H. Gent)

reduces humidity, which may help reduce disease severity. Stripping is accomplished chiefly by the application of chemical desiccants or, rarely, manually. In Australia and New Zealand, where downy and powdery mildew and hop aphids are not present, sheep are sometimes used to graze and remove the basal growth. Care must be used when determining the date and frequency of stripping, because stripping can reduce carbohydrate reserves in the root system and lead to significant yield reductions the following season. Deleterious effects of stripping can be more severe on early-maturing cultivars and plants weakened by soilborne diseases or when little leaf tissue is left at harvest to allow plants to accumulate carbohydrates before winter dormancy.

Cultivation

Cultivation practices vary considerably in different production regions and may include subsoiling only in spring or during alternate years, disking and harrowing, or only shallow (<10 cm) harrowing. Historically, intensive cultivation was common because nutrient requirements were met largely by application of manure and other organic material, necessitating deep disking or plowing for incorporation. Weeds were also controlled by mechanical tillage. However, the frequency and intensity of cultivation has since decreased because of increasing labor costs, exacerbation of Verticillium wilt by cultivation, conversion of hop yards to wider row spacing (i.e., 4.3 to 4.9 m between rows) to facilitate drip irrigation, and the availability of suitable herbicides. With noncultivation, spring pruning and subsequent weed control are accomplished by using herbicides. Research in the United Kingdom has documented that yields obtained under limited cultivation or noncultivation are equal or superior to those obtained by conventional tillage. Another advantage is that noncultivated soils allow farm equipment to enter fields more quickly after rain. Preservation of plants between rows also can provide seasonal habitat for beneficial insects and mites and thereby enhance biological control of some arthropod pests.

Irrigation and Fertilization

Irrigation is required for satisfactory crop yield and quality in most regions of hop production outside Western Europe, where rainfall typically meets crop water demands. The number of irrigation events and the total volume of water needed to produce hop vary widely among different climates and seasons, but in arid climates hop requires approximately 700–800 mm of water during a typical growing season. Various methods of irrigation are utilized, including furrow irrigation, hand-moved sprinklers, overhead sprinklers, and drip. Drip irrigation is typically the most efficient and offers several advantages for crop management since water and nutrients can be metered and delivered directly to the plants, but it requires greater capital expenditures.

In arid climates, irrigation generally begins in mid spring and continues until just prior to the harvest. Irrigation scheduling should aim to balance crop demands, because excessive irrigation can exacerbate several diseases (e.g., black root rot, downy mildew, powdery mildew, and Verticillium wilt) and pollute groundwater by leaching nitrates.

Fertilization is critical to produce hops profitably. Nutrient requirements for hop production vary considerably in the published literature and may differ among cultivars, regions, and seasons. Specific rates and fertility recommendations vary and should be obtained from and verified with local experts and consultants.

Recommended nitrogen rates range from 112 to 300 kg/ha in the United Kingdom, Germany, and the United States and may be reduced if residual soil nitrogen is indicated by soil analyses or if crop debris or manure is applied. Excessive nitrogen fertilization can exacerbate certain diseases (e.g., powdery mildew and Verticillium wilt) and should be avoided.

Phosphorus and potassium fertilization recommendations range from 0 to 100 kg of phosphate per hectare and 0 to 150 kg of potassium per hectare, depending on the region and residual levels in the soil. Recommended fertilization rates for these nutrients have been developed on the basis of soil analyses in several regions and can help to minimize unnecessary applications. Approximately 25–30% of the phosphorus and potassium is found in cones, and these nutrients can be readily returned to hop yards as composted leaf and stem debris after harvest.

Other nutrients that are occasionally required as supplements include magnesium, sulfur, zinc, boron, and molybdenum. Magnesium deficiencies may occur in regions with high rainfall and acidic soils and can be diagnosed with appropriate soil analysis. Symptoms of magnesium deficiency can be diagnostic in some cultivars and more difficult to diagnose in others. Soil can be amended with dolomite to correct magnesium deficiencies. Sulfur deficiencies are common in coarse soils in high-rainfall climates because soluble sulfate anions are readily leached in soil water, but plant-available sulfur is difficult to quantify by soil analysis. Crop needs for sulfur depend on

Fig. 14. Hop plants climbing strings several weeks after training. (Courtesy D. H. Gent)

Fig. 15. Manual training of bines along the string. (Courtesy D. H. Gent)

many factors, including irrigation practices, yield, amount of vegetation removed from the field, fertilization rates of other nutrients (especially nitrogen), and even cultivars. However, scant information is available on sulfur demand for particular hop cultivars. Hop is also very sensitive to zinc deficiencies. Plant roots absorb zinc in its reduced form (Zn^{2+}), and zinc deficiencies are common in alkaline soils because concentrations of zinc in soil solution decrease approximately 30-fold with each unit increase in pH in the range of 5 to 7. Zinc deficiencies can be induced in soils high in phosphorus because of interactions of these nutrients. Soluble zinc salts and complexes can enter plant leaves, and foliar applications of zinc are often used to correct deficiencies. Zinc fertilization can also suppress symptoms of ilarviruses.

Harvesting and Drying

Depending on the cultivar, harvest in the Northern Hemisphere begins in mid to late August and may continue through late September, while in the Southern Hemisphere harvest begins in late February and can continue through early April. Decisions on harvest dates take into account factors such as cone maturity and moisture content, weather and pest threats, and market considerations. Selecting the proper harvest date is critical in achieving optimal yield and quality. Harvesting too late can reduce the aroma and brewing quality of hops because of accelerated oxidation in storage and subsequent loss of volatile aroma compounds and shortened storage ability. Cones damaged by diseases and arthropod pests are particularly sensitive to oxidation during drying and storage, making timely harvest critical. However, harvesting too early reduces yield and flavor constituents of the cone during the current season, and reductions in growth and yield the following season may occur with some cultivars, especially those with low vigor or reserves of carbohydrates. Yield losses the following season from an early harvest can be particularly apparent in early-maturing cultivars, apparently because of disrupted carbohydrate partitioning into the root system.

Hops were once picked by hand; however, automated picking machines are now used throughout much of the world to reduce harvest time and labor costs. With conventional tall trellises, the bines of plants are cut at the base and from the overhead support wires and transported by truck or trailer to stationary picking machines (Fig. 16). A mobile harvester was developed in the United States for picking hops from a 5.5-m trellis, but these harvesters are used on a very limited basis. The plant and string may be cut from the trellis and at the ground by hand or by using specialized equipment. Entire bines are loaded by hand or, less commonly, mechanically into a picking machine that strips and separates cones from the bines, leaves, and stems. With low-trellis systems, mobile picking machines are used to remove cones from plants in situ, leaving most of the bines and crop debris in the field. Cones are then cleaned to remove small pieces of stems and leaves.

Fig. 16. Hop bines waiting to be loaded into a picking machine. (Courtesy D. H. Gent)

After the harvest, crop debris, or "trash," is returned to hop yards or other fields before or after composting. The decision on whether to compost or return the green material to hop yards or other fields is influenced by the pathogens potentially present in the debris (e.g., propagules of *Verticillium* spp. and chasmothecia of *Podosphaera macularis* [Wallr.] U. Braun & S. Takam.) and logistical constraints associated with handling the large volume of material. Significant levels of some nutrients are present in the trash, and returning wastes to agricultural fields can help to reduce fertilizer requirements.

Hop Processing

Immediately after harvest, hops are dried in forced-air kilns heated by methane, propane, or fuel oil. Increases in energy costs have led to new interest in drying hops with ambient or solar-heated air, but these practices are uncommon on most farms. Small quantities of hops can be dried for home use with a food dehydrator, large volumes of ambient air, or other small-scale drying systems. Depending on the drying system, cones are dried at 50–70°C for 4–10 h, reducing the moisture content from approximately 80% to 8–12%. Drying is essential for long-term storage since it reduces spoilage from decay organisms as well as heating and subsequent combustion of stored hops. However, overheating can reduce alpha-acid content and brewing quality, especially when air movement in the kiln is slow, and may cause deterioration of cone color and aroma. Hops damaged by spider mites and other pests are particularly prone to overdrying. After sufficient cooling, the dried hops are compressed into bales and wrapped in burlap or plastic for storage and transit to the end user or processing facilities. The hop storage index is a commonly used analytical method that indicates the potential oxidation of alpha-acids during storage by measuring the percentage of alpha- and beta-acids present in cones. This technique measures the optical density ratio at 275 and 325 nm.

Selected References

Barth, H. J., Klinke, C., and Schmidt, C. 1994. The Hop Atlas. Joh. Barth & Son, Nuremberg.

Brooks, S. N., Horner, C. E., and Likens, S. E. 1961. Hop Production. USDA-ARS Info. Bull. No. 240. U.S. Government Printing. Office, Washington, DC.

Gingrich, G., Hart, J., and Christensen, N. 2000. Hops. Oregon State University Extension Fertilizer Guide FG 79.

Neve, R. A. 1991. Hops. Chapman and Hall, London.

Tomlan, M. A. 1992. Tinged with Gold: Hop Culture in the United States. University of Georgia Press, Atlanta.

Wample, R. L., and Farrar, S. L. 1983. Yield and quality of furrow and trickle irrigated hop (*Humulus lupulus* L.) in Washington State. Agric. Water Manage. 7:457-470.

(Prepared by R. A. Beatson, S. T. Kenny, S. J. Pethybridge, and D. H. Gent)

Cone Uses and Chemistry

The primary purpose of hops as an ingredient of beer is to impart flavor (primarily bitterness) and aroma. Hops mask the sweetness of malt, while at the same time providing aroma to beer. Bittering acids are also important for foam head development and contribute to flavor stability. Hop bittering acids, particularly beta-acids, have antimicrobial activity against gram-positive beer-spoilage microbes (e.g., *Lactobacillus* spp.) and as such contribute to the preservation of beer. Recently, the antimicrobial and medicinal properties of hops have been recognized as having potential applications in areas beyond brewing, as witnessed by the plethora of patents on the topic. Although the application of hops in areas beyond brewing is still in its infancy, its potential has been recognized, and enterprises based on these applications are being developed.

Hop Chemistry

Hops are a raw agricultural commodity, which explains the complexity of their composition. The three primary components of brewing value are the alpha-acids, beta-acids, and essential oils. All these constituents are produced almost exclusively in the lupulin glands of hop cones. Table 1 presents the typical composition of commercial (dry) cone or baled hops, listed in order of decreasing importance from a brewing perspective.

Hop Bitter Acids

By far the most important group of substances in hops are the alpha-acids. Alpha-acids are complex enolic acids with a six-carbon ring structure and several substituent groups (Fig. 17). They are weak dibasic acids, and the acidic properties arise from the dissociation of two enolic hydroxyl groups. There are at least seven alpha-acids, but humulone, cohumulone, and adhumulone make up 98–99% of the alpha-acids in most cultivars (Table 2). Alpha-acids differ from each other only with respect to one acyl side chain (-R group) that may be split off to give a five- or six-carbon fatty acid. Different hop cultivars have varying proportions of these alpha-acids in the total alpha-acid fraction. The development of alpha-acids in the cone begins early in strobile growth. The interactions of environmental factors responsible for alpha-acid levels are extremely complex, and yield factors such as resin gland numbers per cone are affected by environmental conditions (e.g., temperature variability and extremes) before and during flower initiation and pollination for seeded crops.

The most important reaction in brewing chemistry is the isomerization of alpha-acids to iso-alpha-acids (Fig. 17). This isomerization occurs during the boiling of "wort," the sugary liquid produced from heating grist (cracked grains) in water during the early stages of the brewing process. Wort boiling converts the largely insoluble alpha-acids to the more soluble and bitter iso-alpha-acids.

Quantification of the hop bittering acid levels is important and can be conducted by using a variety of methods including the lead conductance value (LCV), spectrophotometry, and column chromatography. The LCV test involves successive addition of small aliquots of a methanolic solution of lead acetate to

Table 1. Typical composition of dried hops

Component	Weight (%)
Alpha-acids	2–15
Beta-acids	1–10
Essential oil	0.5–5
Polyphenols (tannins)	2–5
Oil and fatty acids	trace–25
Wax and steroids	...
Protein	15
Cellulose	40–50
Water	6–10
Chlorophyll	...
Monosaccharides	2
Pectins	2
Minerals (ash)	10

Table 2. Chemistry and relative composition of hop alpha-acids found in cones

Alpha-acid component	Acyl side chain	Alpha-acid fraction (%)
Humulone	$-CH_2CH(CH_3)_2$	35–70
Cohumulone	$-CH(CH_3)_2$	20–55
Adhumulone	$-CH(CH_3)CH_2CH_3$	10–15
Prehumulone	$-(CH_2)_2CH(CH_3)_2$	1–10
Posthumulone	$-CH_2CH_3$	1–5

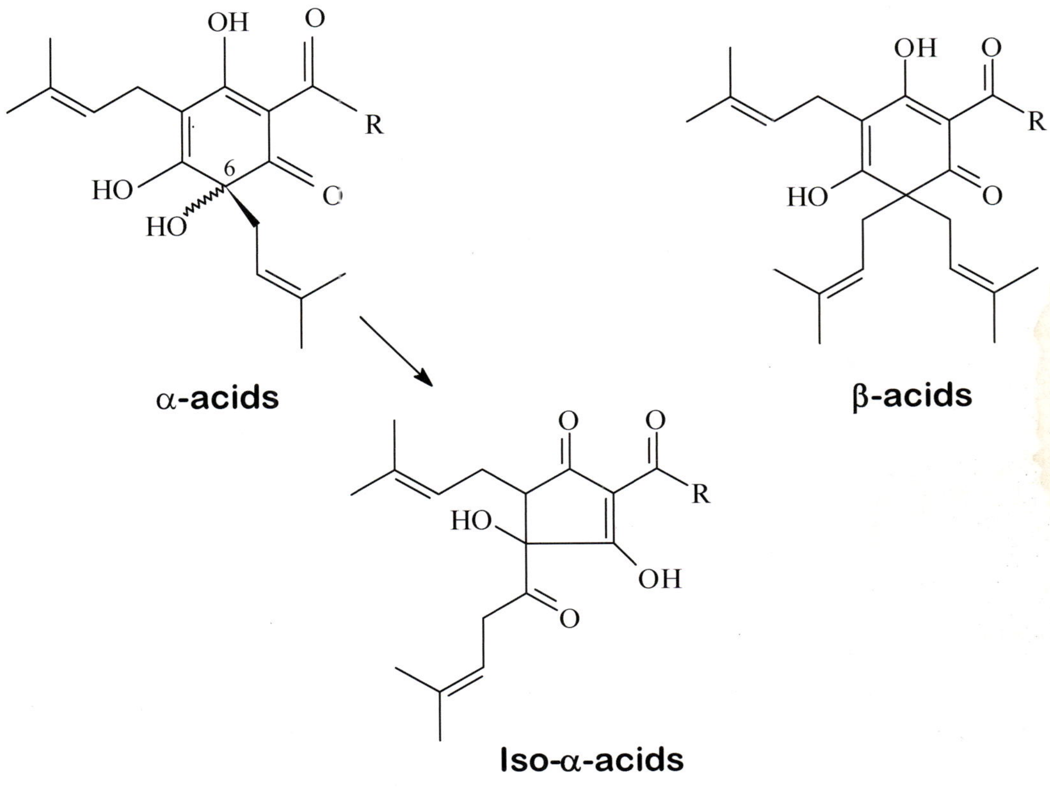

Fig. 17. Structure of alpha-, beta- and iso-alpha-acids. (Courtesy D. Hysert)

a methanolic solution of hop resins. The conductivity of such a solution remains constant while alpha-acids complex with lead ions but increases sharply when lead ions become present in excess. This is represented graphically, and where the two lines intersect is deemed the "end-point." Because of the formation of alpha-acid oxidation products, which can also form lead salts, the final result is quoted as an LCV rather than an actual alpha-acid figure. Advantages of this test are its speed, reproducibility, and relatively low expense. The spectrophotometric method relies upon the principle of pH-regulated ultraviolet light absorption of the hop resins. The maximum absorptions for alpha- and beta-acids are at 325 and 355 nm, respectively, with minimum (background) absorption at 275 nm. Alpha- and beta-acid contents are then calculated by inserting extinction values into appropriate simultaneous ternary equations. This technique is rapid and more accurate than the LCV test. Column chromatography or high-performance liquid chromatography also can be used to accurately measure the proportions of a range of the bittering acids.

Another group of weak organic acids in hops are the beta-acids (Table 3). Beta-acid composition is also characteristic of hop cultivars and varies with the ripening stage. These differ from the alpha-acids in having an isopentenyl side chain in place of the second hydroxyl group at ring position 6 (Fig. 17). The number of analogues is the same as in alpha-acids. These analogues differ only in the nature of the R group (Fig. 17 and Table 3). Beta-acids make only a minor contribution to beer flavor but are important for preservation. The ratio of alpha- to beta-acids varies depending on the stage of hop development and the cultivar, but it typically ranges from 1:1 to 4:1. Cultivars with high levels of beta-acid (>9%) and extremely low levels of alpha-acid (<1%) have been developed for brewing and nonbrewing purposes.

There are a few characteristics of the hop bitter acids that greatly affect their behavior in the brewing process. Bitter acids are soluble in hydrocarbons, alcohols, and chlorinated solvents but are insoluble in water (hydrophobic). At the onset of fermentation, the production of carbon dioxide by yeast causes a drop in pH from approximately 5.2 to 4, reducing the solubility of bitter acids (Table 4). The iso-alpha-acids are more soluble in water than their precursors but are still only slightly water soluble at the pH of beer. Any alpha-acid that has not been isomerized becomes insoluble and is deposited on vessel walls, in the first foam head, or adsorbed to yeast and other sediment. The solubility of the beta-acids in water at pH 5.2 is practically zero. When hops or hop products are boiled in the brew kettle, brewers are attempting to bring these low-solubility substances into solution in the wort. Consequently, high losses are incurred. If starter tubs are used, much of this material is left behind. When skimming is practiced, a considerable amount of alpha-acid is removed. Foam chambers also act as collection points for this precipitated material. Nevertheless, the drop in pH also affects bitter acid solubility during fermentation, and there is a significant loss of iso-alpha-acids at this stage.

Table 3. Chemistry and relative composition of hop beta-acids found in cones

Beta-acid component	Acyl side chain	Beta-acid fraction (%)
Lupulone	$-CH_2CH(CH_3)_2$	30–55
Colupulone	$-CH(CH_3)_2$	20–55
Adlupulone	$-CH(CH_3)CH_2CH_3$	5–10
Prelupulone	$-(CH_2)_2CH(CH_3)2$	1–3
Postlupulone	$-CH_2CH_3$	<1

Table 4. Aqueous solubility of hop bitter acids

pH	Humulone (ppm)	Lupulone (ppm)
7.0	...	48
6.2	740	16
5.6	244	8
5.2	84	Nondetectable
4.6	20	Nondetectable
4.0	8	Nondetectable

Hop Volatile Oil

Another major group of hop constituents are the volatiles or essential oils. This group is generally thought to be the source of hoppy aroma in beer, but obtaining and controlling this trait is one of the most elusive problems in brewing. Hop oil is a very complex mixture, probably containing over 300 different chemical entities made up of hydrocarbons and oxygenated and sulfur-containing compounds. These hydrocarbons can be aliphatics, monoterpenes, or sesquiterpenes. The ratios of specific volatile oils (farnesene, humulene, and β-caryophyllene) as well as the content of specific "floral" or "fruity" volatile oils define brewing quality. The proportions of these entities in the oil vary among cultivars and batches. Levels of volatile oils generally increase logarithmically with cone ripening. Loss of volatile oils is of particular concern with the use of aroma hops. Loss can occur in storage through oxidation, polymerization, or resinification of components, machine picking, drying, and poor baling and pelleting techniques.

Hop Processing and Products

Prior to the late 1930s, there was only one form in which hops were used in brewing in North America, namely, the standard American bale. The bale is wrapped in heavy cloth (burlap) and weighs approximately 90 kg. Currently, there are more than 25 hop products from which brewers can choose, although not all of these are available in all countries (Tables 5 and 6). The reasons for processed hop products, and thus the hop processing industry, is brewer demand for lower costs, greater quality control, environmental and product safety, and production of specialty beers.

Over the last several years, sales of hops and hop products worldwide have been shifting from traditional products (pellets and non-isomerized hop extracts) to advanced hop products (Tables 5 and 6). Although the total volume of advanced hop

Table 5. Hop products available in 2006 in addition to baled hops

Pellets	Non-isomerized extracts	Isomerized extracts	Aromatic oils
Regular (Type 90)	Pure resin	Isomerized kettle extract (IKE)	Whole hop oil
Enriched (Type 45)	Diluted resin	Isomerized post fermentation extracts	Oil-rich CO_2 extract
Stabilized	Ethanol extracted	Normal isomerized extracts	Late hop essence
Isomerized	Carbon dioxide extracted (super- and sub-critical)	Reduced isomerized extracts (Rho [Dihydro], Tetra- and Hexa-hydro)	Pure hop aroma (PHA)

Table 6. Worldwide distribution of hops and hop products sold in 2005

Hop product class	% Alpha-acid sold
Baled hops	1
Pellets	62
Extract	
Traditional hop extract products	29
Advanced hop extract products	
Tetra	3.8
Isomerized kettle extract (IKE)	1.8
Rho	1.6
Iso	0.72
Hexa	0.08

products sold in 2005 accounts for only 8% of the total hop alpha-acids used worldwide, this amount is nevertheless important and reflects the increasing use of advanced hop products for the purposes mentioned above.

Selected Reference

Peacock, V. 1998. Fundamentals of hop chemistry. Master Brewers Assoc. Am. Tech. Q. 35:4-8.

(Prepared by D. Hysert)

Hop Cultivars and Breeding

Hop cultivars can be divided into two broad types, which are distinguished by their use during the brewing process. Cultivars used for bittering contain high levels of alpha-acids, have a high alpha- to beta-acid ratio (3:1), and have a lower humulone to cohumulone ratio (Table 7). The storage life of some bittering hops is generally poor, ripening is in mid to late season, aroma content is typically considered unimportant, and overall yield is generally good. However, bittering power is a combination of genotypic and environmental factors. Breeding programs worldwide have been extraordinarily successful at producing high alpha-acid cultivars. These have become known as "super-alpha" cultivars and are defined as having alpha-acid levels above 15%, with some cultivars having alpha-acid levels approaching 20%.

Aroma, or "noble," hops are produced to enhance beer flavor (Table 7). Aroma cultivars have low alpha-acid levels (4–9%), low alpha- to beta-acid ratio (1:1), and high essential oil levels (>3 ml of oil/100 mg of hop cone tissue). However, oil content and ratios of specific compounds vary among cultivars and are affected by harvest date. Aroma hop cultivars generally have lower yields than bittering cultivars, but the price paid per unit is typically higher. While the alpha-acid content of both bittering and aroma hops declines in storage, the bittering potential of the latter declines more slowly. This occurs because aroma hops have proportionately more beta-acids that form water-soluble bitter oxidation products and compensate somewhat for the loss in alpha-acids. The percentage of myrcene and humulene in the volatile oil is inversely related and differs between these two groups of cultivars. Myrcene as a percentage of total volatile oils is generally 20–45% in aroma hops compared with 34–65% in bittering hops. In addition, aroma hops typically have a higher humulene to caryophyllene ratio (>2.5) while that of bittering hops is generally lower.

Historically, the primary objective of hop breeding programs has been to increase the yield or characteristics associated with either bittering (high alpha-acids) or aroma (unique volatile oil profiles) cultivars. Other factors considered during breeding are resistance to diseases and arthropod pests, cone structure and color, and a variety of agronomic characteristics including ripening date, bine twining ability, and internodes length. Early breeding likely involved mass selection of superior genotypes (landraces) from mixed populations of cultivated or feral plants and selection of seedlings resulting from open pollination. Examples of cultivars arising from mass selection and subsequent propagation include Hallertauer Mittelfrüh, Hersbrucker, and Saazer. Notably, the aroma cultivar Fuggle is thought to have been selected from a chance seedling collected by Richard Fuggle in the United Kingdom.

Directed breeding programs, developed in the early twentieth century, relied on hybridization of superior females via open pollination and later designed crosses. Currently active hop breeding programs rely largely on pedigree breeding and male test-crosses to aid in selecting male parental lines. However, breeding objectives have expanded beyond agronomic and brewing applications as potential alternative markets for hop products have arisen. Examples of other breeding objectives now include increased production of non-brewing chemicals (e.g., xanthohumol for anti-cancer properties, 8-prenylnaringenin for estrogenic benefits, and elevated beta-acid levels for antimicrobial applications) and development of high-yielding, low-trellis cultivars and ornamental cultivars.

Because hop is monoecious, emasculated crosses are not necessary for hybridization. A number of cultivars have arisen from open pollination of specific female accessions by nearby male lines; Brewers Gold, Cascade, and Galena are three of the more important cultivars developed in this fashion. Most cultivars are developed via controlled crosses between a single male and single female line. During most years, male inflorescences begin shedding pollen immediately prior to female receptivity and shed pollen for several weeks after that point. Pruning and training male plants earlier than female plants enhances differences in timing of inflorescence development. An efficient means to capture sufficient pollen for crosses is to collect and hang male lateral branches bearing active inflorescences over wax paper for several hours.

Immediately prior to female inflorescence receptivity (early burr stage), an entire lateral branch of a specific cultivar to be used as a female parent can be defoliated and enclosed within a waxed, lightweight bag. Once female inflorescence receptivity has been reached, an opening is cut on the top of the bag, pollen is poured into it, and the bag is resealed and shaken to evenly spread the pollen across the flowers. If lightweight white paper bread bags are used, one can leave the bags on the lateral branch until seed maturity, which typically occurs 8 weeks after pollination.

Hop plants are usually diploid (n = 20); however, advances in breeding technologies have produced triploid cultivars with higher alpha-acid contents, yields, and infertility (low seed levels). Triploid breeding programs are active in New Zealand and Australia, where bittering triploid cultivars have been adopted widely. Triploid aroma cultivars also are used in commercial production in the United States and Europe, although breeding efforts in the United States and Europe have shifted away from triploids because of yield instability and the inability to use triploid lines in subsequent crosses. The use of infertile triploid cultivars is considered advantageous in regions where feral male plants reside because the presence of seed adds substantial weight to dried hops, increasing storage and freight costs. Also, seeds are often undesirable to breweries because oxidation of the seed fat is responsible for the production of impurities, adversely affecting beer color and flavor. Triploids are produced by crossing a tetraploid female parent, in which the chromosome number has been doubled with a mutagenic agent (e.g., colchicine), with a diploid male. Mutagenesis also has been used to generate genetically stable diploid plants with superior agronomic characteristics. This process involves exposure of axillary buds to mutagenic agents such as ethyl methanesulfonate.

A range of molecular-based techniques, such as marker-assisted selection, isolation and cloning of quantitative trait loci, and characterization of genes involved in pest defense pathways, are also being explored to advance hop breeding worldwide. Identity and purity of some cultivars can be verified by using a variety of DNA fingerprinting techniques. Genetic transformation has been used experimentally to introduce desirable characteristics (e.g., enhanced disease resistance) into known hop cultivars without changing brewing quality profiles. In other efforts, the valerophenone synthase gene responsible for the production of hop resin biosynthesis has been isolated and is being used to improve the contents or components of bioactive substances. Although genetic transformation of hop is being researched actively in Europe and Japan, genetically engineered hop cultivars have not yet been released for commercial production.

Table 7. Country of origin, usage and characteristics of selected hop cultivars and accessions

Country/Cultivar[a]	Pedigree[b]	Usage	Disease susceptibility[c]			Chemical characteristics[e]							
			PM	DM	VW[d]	Alpha	Beta	CoH	HSI	Myr	Car	Far	Hum
Australia													
Pride of Ringwood	Pride of Kent/OP//OP	Bittering	U	S	T	6.1	5.0	37.5	0.3	45.9	7.2	0.2	25.5
Czech Republic													
Saazer	Old Czech landrace	Aroma	S	T	U	4.2	3.6	23.0	0.3	33.0	7.5	13.3	24.7
Saazer 36	Clonal selection from Saazer	Aroma	S	T	T	5.1	4.2	22.3	0.3	49.6	4.6	12.8	14.1
Sirem	Mass selection from Saazer	Aroma	S	T	U	3.5	3.1	23.0	0.3	27.0	6.1	15.7	22.5
Zlatan	Clonal selection from Saazer	Aroma	S	S	U	5.2	4.3	21.0	U	U	U	U	U
Denmark													
Nordgard 1478	Nordgard 978//Bramling Cross/OP	Aroma	U	U	U	7.6	3.4	29.3	0.3	47.3	10.5	1.7	26.8
United Kingdom													
Alliance	WGV/Vert-R breeding line	Aroma	S	U	T	6.1	2.4	29.9	0.3	36.5	11.9	2.2	34.3
Brewers Gold	BB1/OP	Bittering	S	T	R*	9.3	4.8	39.5	0.3	66.7	6.5	0.1	11.3
Bullion	BB1/OP	Bittering	S	T	R	10.7	5.2	38.4	0.3	63.3	8.8	0.1	12.5
Density	Bullion/OP	Bittering	S	S	R	5.3	3.3	36.7	0.3	57.3	6.5	0.3	16.8
East Kent Golding	Old English cultivar	Aroma	S	S	T	5.1	2.3	28.3	0.3	31.7	25.3	U	29.6
Eastwell Golding	Old English cultivar	Aroma	S	S	T	6.0	3.0	30.0	U	50.0	8.0	U	25.0
First Gold	WGV/23_77_56M	Bittering	R	S	T	7.2	3.0	31.5	U	U	U	U	U
Fuggle	Old English cultivar	Aroma	S	R	S	5.0	2.5	26.9	0.3	43.4	9.1	4.3	26.6
Herald	Wye34_73_11//Omega/dwarf	Bittering	R	T	T	12.0	5.2	37.0	U	U	U	U	U
Keyworth Midseason	Y90/OP	Aroma	U	U	U	7.5	3.1	46.5	0.3	59.2	7.6	U	10.0
Northern Brewer	Canterbury Golding//BG/male	Bittering	S	R	T*	10.2	3.9	23.2	0.3	55.0	8.8	0.1	20.9
Phoenix	Yeoman/Wye38_77_14M	Bittering	R	S	R	13.0	4.0	26.0	U	U	U	U	U
Pioneer	Wye34_73_11//Omega/dwarf	Bittering	T	T	U	9.0	3.8	37.0	U	U	U	U	U
Progress	WGV/OB79	Bittering	S	T	T	6.7	2.7	27.8	0.3	32.8	12.3	0.3	37.6
Tolhurst	Old English Variety	Aroma	U	U	U	2.4	2.6	32.0	0.3	45.1	6.0	8.5	17.3
Tutsham	Clonal selection of Golding	Aroma	S	S	S	U	U	U	U	U	U	U	U
WGV	Whitsbred Golding Variety	Aroma	T	S	T	7.7	2.5	32.5	0.2	42.2	9.2	1.4	26.7
Wye Challenger	Zattler/OP//NB/Wye22_56	Bittering	R	R	S	7.8	4.8	23.8	0.3	56.1	5.7	0.9	17.0
Wye Northdown	NB/OP	Bittering	S	T	S	8.5	5.2	23.6	0.3	56.2	9.7	0.6	21.3
Wye Omega	Challenger/English male	Bittering	S	T	R	8.8	3.1	28.0	0.3	54.2	10.6	U	14.1
Wye Saxon	Svalof//Bramling Cross/Wye1_63_42	Bittering	R	R	T	8.2	4.3	19.4	0.3	49.4	6.6	6.5	13.8
Wye Target	NB/Wye22_56//Eastw. Golding/OB79	Bittering	R	S	R	12.2	5.1	34.7	0.3	65.0	5.1	0.1	9.7
Wye Viking	Svalof//Bramling Cross/Wye1_63_42	Bittering	R	R	S	7.8	4.2	24.0	0.3	47.0	5.9	9.3	10.5
Wye Yeoman	Wye43_69_17/Wye25_68_173	Bittering	R	R	R	13.3	4.6	23.9	0.3	46.5	20.1	0.1	20.8
Wye Zenith	Wye25_68_22/OP	Bittering	R	R	S	10.1	3.2	21.8	0.3	49.5	21.2	0.1	20.5
Former Yugoslavia													
Backa	Mass selection of Bavarian hop	Aroma	S	S	U	4.2	6.0	24.3	0.2	49.5	14.7	0.1	24.6
Dunav	NB/OP	Bittering	S	T	T	6.9	3.6	29.8	0.3	59.9	5.8	2.5	18.2
Neoplanta	NB//Savinski Golding/wild hop	Aroma	S	R	T	9.2	3.9	35.9	0.3	50.2	8.6	5.9	19.5
Vojvodina	NB/SX-502	Bittering	U	R	S	8.6	3.6	29.4	0.3	66.7	4.9	0.6	13.0
France													
Strisselspälter	Possibly from Hersbrucker Spält	Aroma	U	T	T	4.7	3.2	27.4	0.3	47.4	10.7	U	27.4
Tardif de Bourgogne	French landrace	Aroma	U	R	U	4.5	4.9	19.9	0.2	40.3	5.8	U	14.1
Germany													
Hall. Magnum	Galena/(U)	Bittering	S	T	T	14.7	6.5	26.0	0.2	72.6	10.5	U	10.6
Hall. Mittelfrüh	Mass selection of Bavarian hop	Aroma	T	S	S	4.8	4.3	19.4	0.3	33.4	10.6	0.3	37.7
Hall. Taurus	82_39_37/85_54_15M	Bittering	S	T	T	U	U	U	U	U	U	U	U
Hall. Tradition	Hall. Gold/(U)	Aroma	T	T	T	6.6	5.1	26.0	0.2	66.9	9.4	U	13.9
Hall. Gold	Hüller Anfang/8lt9//NB/54_33_11M	Aroma	T	T	T	8.0	6.4	19.0	0.2	67.8	7.9	U	13.3
Herkules	Hall. Taurus/93_09_41	Bittering	R	T	T	U	U	U	U	U	U	U	U
Hersbrucker Pure	73_10_16/(U)	Aroma	T	R	R*	5.3	2.8	23.5	0.3	38.5	24.4	U	9.9
Hüller Bitterer	NB/(2N=4X=40, Hüller)	Bittering	T	R	S*	8.5	4.7	27.9	0.3	54.7	7.7	0.1	7.9
Hüller Fortschritt	Hüller Anfang//8lt93o45	Aroma	T	R	S	U	U	U	U	U	U	U	U
Merkur	Hall. Magnum/81_08_13	Bittering	R	T	T	U	U	U	U	U	U	U	U
Opal	Hall. Gold/KuII	Aroma	T	T	T	U	U	U	U	U	U	U	U
Orion	Perle/70_10_15	Aroma	T	T	T*	8.8	5.3	34.5	0.2	48.0	19.6	U	22.7
Perle	NB/(U)	Aroma	S	R	R*	10.0	3.9	27.7	0.3	43.0	11.7	0.1	29.3
Saphir	83_17_20/80_56_06	Aroma	T	T	T	U	U	U	U	U	U	U	U
Spalter	Mass selection from Saazer Hop	Aroma	U	R	U*	5.0	4.0	22.7	0.3	48.4	5.2	9.8	16.6
Spalter Select	76_18_80/71_18_145	Aroma	T	R	R	4.1	4.5	25.0	0.2	67.9	2.1	8.0	5.4
Tettnanger	Old German landrace	Aroma	U	T	T	4.3	3.6	23.5	0.3	39.5	6.9	11.9	22.3
Wurtemberger	Mass selection from Bavarian hop	Aroma	U	T	T	5.4	2.9	31.0	0.2	63.7	5.1	3.3	14.3
Japan													
Golden Star	Mutant from Shinshuwase	Aroma	U	R	U	5.5	4.5	50.2	0.3	53.6	5.7	U	13.6
Eastern Gold	C76-64-17/OB79	Bittering	R	T	U	11.4	5.4	27.0	0.2	46.1	7.2	2.8	16.7

Table 7 ***(continued)***

Country/Cultivar[a]	Pedigree[b]	Usage	Disease susceptibility[c] PM	DM	VW[d]	Chemical characteristics[e] Alpha	Beta	CoH	HSI	Myr	Car	Far	Hum
Kirin I	Shinshuwase's clonal selection	Aroma	U	T	R	U	U	U	U	U	U	U	U
Kirin II	Saazer//White Bine/OP	Bittering	U	R	U	6.6	5.6	47.0	0.3	51.4	15.4	U	13.6
Kitamidori	C79-27-01//Kirin II/OB79	Bittering	U	S	U	5.0	5.1	22.5	0.2	38.9	7.5	6.0	24.8
Shinshuwase	Saazer/White Bine	Aroma	U	T	R	5.6	4.6	50.1	0.3	58.1	12.0	U	11.6
Toyomidori	NB/OB79	Bittering	U	S	T	12.4	5.2	44.5	0.2	68.4	10.4	U	5.9
New Zealand													
Calicross	Late Cluster/Fuggle/OP	Aroma	U	S	U	7.0	5.5	37.9	0.3	62.4	10.7	0.2	15.4
First Choice	Late Cluster/Fuggle/OP	Aroma	U	S	S	5.8	4.8	39.0	0.3	73.8	1.4	6.9	0.7
Smooth Cone	Late Cluster/OP	Aroma	U	S	R	8.1	4.2	29.8	0.3	55.2	6.2	0.4	20.7
Sticklebract	First Choice(2n=4X=40)/OP	Aroma	U	T	T	9.9	4.7	42.1	0.3	60.4	4.0	4.3	8.5
SuperAlpha	Smooth Cone(2n=4X=40)/OP	Bittering	U	T	T	10.7	5.4	35.8	0.3	60.1	20.0	0.0	16.7
Russia													
Serebrianka	Wild Siberian hop selection	Aroma	R	T	U	3.5	3.2	23.2	0.3	28.2	8.2	12.0	27.0
Poland													
Lubelski	Probably Saazer ancestry	Bittering	U	T	U	4.8	4.1	22.5	0.3	46.0	5.6	11.9	19.6
Slovenia													
Ahil	BG/3_3	Bittering	U	T	T	10.8	4.3	25.3	0.3	62.6	4.4	10.1	8.5
Apolon	BG/3_3	Bittering	U	T	T	10.7	3.8	25.4	0.3	58.4	5.2	12.7	11.7
Atlas	BG/3_3	Bittering	U	T	T	9.7	3.9	36.1	0.3	58.6	4.4	13.5	8.8
Aurora	NB/TG	Bittering	U	T	T	10.9	4.2	21.8	0.3	49.5	5.8	4.8	18.9
Blisk	Atlas(2n=4x=40)/1_9	Bittering	U	R	T	11.2	4.0	33.6	0.3	57.1	3.9	16.2	10.0
Bobek	NB/TG	Bittering	U	T	R	8.3	5.1	26.0	0.3	62.8	3.9	2.9	12.8
Buket	NB/2_137	Bittering	U	R	R	11.0	4.9	24.0	0.3	58.4	5.4	4.3	15.4
Cekin	Aurora/3_3(2n=4x=40)	Bittering	U	T	U	6.2	2.1	22.3	0.3	49.6	6.3	9.0	18.4
Celeia	Savinjski Golding(2n=4x=40)/105_58	Aroma	U	T	U	5.7	3.9	23.0	0.2	48.0	7.6	6.8	18.3
Cerera	Savinjski Golding(2n=4x=40)/105_58	Aroma	U	T	U	5.7	3.6	23.5	0.3	58.1	6.0	3.2	13.3
Cicero	Aurora/3_3(2n=4x=40)	Bittering	U	T	U	6.7	2.4	28.0	0.3	53.8	6.4	2.7	15.0
Savinjski Golding	Fuggle clonal selection	Aroma	U	R	S	5.3	2.8	28.3	0.3	46.7	8.6	4.1	25.6
South Africa													
Southern Brewer	Fuggle/OP//Fuggle/OP	Aroma	U	R	R	9.6	4.2	39.7	0.3	53.1	6.4	6.3	14.2
United States													
Cascade	Fuggle///Serebrianka//Fuggle/OP///OP	Aroma	R	T	U*	6.2	5.4	37.1	0.3	68.1	3.6	3.9	9.3
Centennial	BG///BG///Fuggle//Fuggle/OP////63015M	Bittering	U	T	T	12.1	4.2	28.6	0.3	77.7	6.1	0.1	5.2
Chinook	Peltham Golding//BG/wild hop	Bittering	U	T	R	13.7	3.8	33.7	0.3	49.4	12.2	0.3	13.2
Columbia	Fuggle (2n=4x=40)//Fuggle/OP	Aroma	S	R	S	8.4	4.0	39.6	0.3	56.1	7.2	3.9	16.5
Comet	Sunshine/OP//Utah 524-2	Bittering	S	S	U	11.2	4.4	41.0	0.3	67.9	15.9	0.0	0.7
Crystal	Hall. Mittelfrüe(2n=4x=40)/21381M	Aroma	T	S	R	5.9	6.8	20.7	0.3	58.2	5.7	0.2	20.0
Galena	BG/OP	Bittering	S	S	U	11.2	7.1	39.0	0.3	53.9	6.7	0.1	10.0
Horizon	65009/64035M	Bittering	S	S	U	5.4	4.3	43.3	0.3	58.1	13.9	0.4	13.4
Late Cluster	Unknown European/Wild American	Aroma	S	S	U	6.0	4.2	40.0	0.3	40.3	7.8	0.3	18.9
Liberty	Hall. Mittelfrüe(2n=4x=40)/64035M	Aroma	T	T	U	5.9	3.8	22.8	0.3	46.6	11.0	0.2	28.7
Mount Hood	Hall. Mittelfrüe(2n=4x=40)/19058M	Aroma	R	S	U	7.7	7.1	22.7	0.3	47.1	12.4	0.2	26.2
Newport	Magnum/58111M	Bittering	R	R	U	14.5	7.8	37.8	0.2	50.0	7.3	0.1	14.0
Nugget	65009/63015M	Bittering	R	S	U	13.3	4.5	25.6	0.3	55.5	9.0	0.2	17.9
Olympic	BG//BG/OP///Fuggle//Fuggle/OP////64028M	Bittering	U	T	R	12.2	5.1	31.3	0.3	67.5	10.4	2.2	7.0
Sterling	Saazer 38///Cascade//65009/64035M	Aroma	T	T	U	8.1	4.7	23.6	U	49.8	5.0	14.5	17.0
U.S. Tettnanger	Clonal selection from Fuggle	Aroma	S	R	S	5.7	2.9	26.7	0.3	54.6	6.6	3.7	19.7
USDA 21055	Comet/65009M	Bittering	S	S	U	14.4	4.7	46.0	0.3	56.1	5.7	7.2	9.2
USDA 65009	BG//Fuggle/Colorado 2-1	Bittering	U	T	U	10.5	8.1	38.0	U	50.0	11.0	U	17.0
Willamette	Fuggle (2n=4x=40)//Fuggle/OP	Aroma	S	T	S	6.1	3.6	33.7	0.3	49.3	7.7	5.0	22.5

[a] Only cultivars tested at the USDA-ARS hop research farm in Corvallis, Oregon, for at least 3 years are presented.

[b] Abbreviations used for some of the more common lines are Brewers Gold (BG), Hallertaür (Hall.), Northern Brewer (NB), Whitbred Golding variety (WGV), and open pollinated crosses (OP).

[c] PM = powdery mildew; DM = downy mildew; VW = Verticillium wilt; U = susceptibility information is not available for these cultivars; R = resistant; T = tolerant; and S = susceptible. Disease reactions are based on published information from multiple sources, and may not be consistent in all environments. Resistant = the ability of the cultivar to prevent or impede disease development; susceptible = prone to developing disease when infected by the pathogen; tolerance = ability to endure disease without serious damage or yield loss. Cultivars are listed as resistant to powdery mildew if they possess known resistance to at least one race of *Podosphaera macularis*.

[d] * = Variable reaction to Verticillium wilt depending on location of evaluation.

[e] Chemical data reported are averages of at least 3 years of cultivar performance in Oregon, U.S. Alpha = alpha-acid (%); Beta = beta-acid (%); CoH = cohumulone (%); HSI = hop storage index; Myr = myrcene (%); Car = caryophyllene (%); Far = farnesene (%); and Hum = humulene (%).

Selected References

Burgess, A. H. 1964. Hops: Botany, Cultivation and Utilization. World Crop Books, Interscience Publication, New York.

Čerenak, A., Jakše, J., and Javornik, B. 2004. Identification and differentiation of hop varieties using simple sequence repeat markers. J. Am. Soc. Brew. Chem. 62:1-7.

Henning, J. A. 2006. The breeding of hop. Pages 102-122 in: Brewing: New Technologies. C. W. Bamforth, ed. CRC Press, Boca Raton, FL.

Moir, M. 2000. Hops—A millennium review. J. Am. Soc. Brew. Chem. 58:131-146.

Seefelder, S., Ehrmaier, H., Schweizer, G., and Seigner, E. 2000. Genetic diversity and phylogenetic relationships among accessions of hop, *Humulus lupulus*, as determined by amplified fragment length polymorphism and fingerprinting compared with pedigree data. Plant Breed. 119:257-263.

Sustar-Vozlic, J., and Javornik, B. 1999. Genetic relationships in cultivars of hop, *Humulus lupulus* L., determined by RAPD analysis. Plant Breed. 118:175-181.

Townsend, M. S., and Henning, J. A. 2005. Potential heterotic groups in hop as determined by AFLP analysis. Crop Sci. 45:1901-1907.

(Prepared by J. A. Henning, S. J. Pethybridge, and D. H. Gent)

Part I. Infectious Diseases

Diseases Caused by Fungi and Oomycetes

Alternaria Cone Disorder

Alternaria cone disorder is generally of minor importance, but it can occasionally damage cones and reduce crop quality. The disease is most common on late-maturing cultivars exposed to wind injury. Alternaria cone disorder has been reported from Europe, Australia, and the United States, but the pathogen is widespread in most hop yards and agricultural systems worldwide.

Symptoms

Disease symptoms vary depending on the degree of mechanical injury to cones. On undamaged cones, the symptoms appear first on the tips of bracteoles of developing or mature cones as a nondescript, light brown discoloration and necrosis (Fig. 18). Bracts may remain green, giving cones a striped or variegated appearance. When cones are damaged by wind or other mechanical abrasion, disease symptoms may appear on both bracteoles and bracts. The disease can progress rapidly, and the necrotic tissues become dark brown and may be confused with damage caused by powdery or downy mildew. Affected bracts and bracteoles may display a slight distortion or shriveling of the diseased tissues. Premature senescence of cones has been attributed to the disease. Damage from Alternaria cone disorder may be limited to one or a few bracts and bracteoles, but in severe cases entire cones may become discolored and necrotic.

Fig. 18. Cone affected by *Alternaria alternata* displaying reddish-brown discoloration of bracts and bracteoles. (Courtesy S. J. Pethybridge)

Causal Organism

Alternaria cone disorder is caused by *Alternaria alternata* (Fr.:Fr.) Keissl. On potato carrot agar, young cultures of *A. alternata* are subhyaline and sporulate profusely by surface and aerial hyphae. Concentric rings of sporulation become apparent within 3–5 days. Primary conidiophores are simple or geniculate with one to four septa and are olive brown. Conidia are greenish brown and slightly beaked and have numerous transverse and longitudinal septa. The conidia may vary in size depending on the substrate, ranging from 7 to 18 × 11 to 65 µm, and are elliptical to ovid. The first conidia produced on conidiophores tend to be larger than conidia borne at the distal end of the chain.

Disease Cycle and Epidemiology

A. alternata is ubiquitous in nature, a common inhabitant of the hop leaf surface, and prevalent in the air of hop yards. Certain strains of the fungus may be phytopathogenic, clinical pathogens, or saprophytic, and they can be isolated from senescent tissue of nearly any plant. However, pathogenic variability and saprophytic fitness of isolates of *A. alternata* on hop have not been investigated. Alternaria cone disorder is a disease primarily of senescent cones compromised by mechanical injury. Severe epidemics are often associated with wind injury accompanied by high humidity or extend periods of dew. Temperatures greater than 18°C during wetting events favor spore germination and penetration on other hosts. *A. alternata* survive and overwinter in and on crop debris, decaying organic matter, and as weak pathogens on many plants.

Management

Mechanical injury of burrs and cones during the application of pesticides and field operations should be avoided. Cultural practices that reduce the duration of wetness on cones by promoting air circulation in the canopy and timing irrigation appropriately may reduce disease severity. Certain fungicides applied to control powdery and downy mildew likely provide some suppression of Alternaria cone disorder if applied near harvest, but there are no reports of formal evaluation trials.

Selected References

Darby, P. 1988. *Alternaria alternata* infection of hop (*Humulus lupulus*) cones. Trans. Brit. Mycol. Soc. 90:650-653.

Góra, A., Skórska, C., Sitkowska, J., Prażmo, Z., Krysińska-Traczyk, E., Urbanowicz, B., and Dutkiewicz, J. 2004. Exposure of hop growers to bioaerosols. Ann. Agric. Environ. Med. 11:129-138.

Pethybridge, S. J., Hay, F. S., Wilson, C. R., Sherriff, L. J., and Leggett, G. W. 2001. First report of infection of hop cones by *Alternaria alternata* in Australia. Plant Dis. 85:804.

(Prepared by D. H. Gent)

Armillaria Root Rot

Armillaria root rot, also known as shoestring root rot, is a common disease of numerous forest and orchard trees, shrubs, and vines in temperate and tropical regions of the world. Little is known about the etiology and epidemiology of the disease on hop, but it appears to be of minor importance under standard production practices.

Symptoms

The initial disease symptom is wilting of the plants. Plaster-white sheets of mycelium from the pathogen are apparent under the bark of infected bines near the soil surface. Thread-like, reddish-brown to black mycelial strands (rhizomorphs) 1–5 mm in diameter may form but are often absent on hop. As the disease progresses, the crown may display a powdery rot. Plants of any age may be killed by Armillaria root rot.

Causal Organism

The first report of Armillaria root rot on hop was in 1937, and it was attributed to *Armillaria mellea* (Vahl:Fr.) P. Kumm. sensu lato. Since then, at least 10 other *Armillaria* spp. have been described, and *A. mellea* is now thought to be primarily a pathogen of hardwoods. It is unclear which species are pathogenic to hop, although the disease is known to be caused by an *Armillaria* sp. other than *A. mellea* in the Southern Hemisphere.

Cultural characteristics of *A. mellea* vary widely, but the basidiocarps are diagnostic when present. These mushrooms are produced in the fall to early winter, range in diameter from 4 to 28 cm, and vary in color from yellow to light or dark brown. Mycelia of *A. mellea* are luminescent.

Disease Cycle and Epidemiology

The disease cycle of Armillaria root rot on hop has not been studied extensively, but the occurrence of the disease is invariably traceable to a nearby stump or hedgerow inoculum source. The pathogen requires woody plant material to survive in the soil but may spread through infected roots or by the growth of rhizomorphs in the soil. An infection of hop by *A. mellea* in the United Kingdom was attributed to an oak stump inoculum source nearly 7 m away. The importance of wind or splash dispersal of basidiospores in the disease cycle is uncertain but is likely negligible for the disease on hop. The pathogen may be disseminated on contaminated equipment and in surface waters, but the contribution of these sources of inoculum for Armillaria root rot on hop has not been quantified.

Management

No control measure provides complete suppression of Armillaria root rot, but the disease is rarely a concern on hop. New yards should not be planted after susceptible tree crops. If a hop yard must be established following a tree crop, all roots and stumps should be removed and destroyed if the disease was present. Fumigation with methyl bromide and chloropicrin suppresses the disease in other crops but is likely uneconomical and unnecessary for hop.

Selected References

Burgess, A. H. 1964. Hops: Botany, Cultivation, and Utilization. Leonard Hill, London.

Salmon, E. S., and Ware, W. M. 1937. The honey fungus (*Armillaria mellea*) attacking fruit trees and hops; with observations on *Pholiota squarrosa* in cherry orchards. J. S. E. Agric. Coll. Wye. 40:18-26.

(Prepared by D. H. Gent)

Ascochyta Leaf Spot

Ascochyta leaf spot has been reported on hop in India, continental Europe, and Asia. It is apparently of minor economic importance in most regions of hop production.

Symptoms

Disease symptoms appear initially as irregularly shaped lesions that begin on the upper surfaces of leaves. Lesions may also form on stems. On leaves, lesions are 1–2 cm in diameter and have a dark grayish margin. Under conditions favorable to the disease, lesions often coalesce and cause premature senescence and defoliation. Minute, dark black specks (pycnidia) are visible in lesions under low magnification.

Causal Organism

Ascochyta leaf spot is caused by *Ascochyta humuli* Kabát & Bubák (syns. *A. humuliphila* Melnik and *Phloeospora humuli* Westend). Pycnidia form predominately in leaves. They are dark brown to black, spherical (60–195 μm in diameter), and possess a small papillate ostiole surrounded by small dark cells. Conidia are hyaline and ovoid to oblong-ellipsoidal, measuring 4.9–8.2 × 1.7–3.3 μm.

Disease Cycle and Epidemiology

The disease cycle and epidemiology of Ascochyta leaf spot have not been investigated. Presumably, the pathogen overwinters on crop debris or in soil and is dispersed by splashing water and wind in the same way that other *Ascochyta* spp. are dispersed on related hosts. Disease symptoms are reported to develop after rain events.

Management

Control measures have not been reported. Certain broad-spectrum fungicides applied for control of downy and powdery mildews may be effective against Ascochyta leaf spot, but there are no reports of formal evaluation trials.

Selected Reference

Singh, G., Singh, R. N., and Bhandari, T. P. S. 1985. Ascochyta leaf spots on fig and hop. New records for India. Indian J. Mycol. Plant Pathol. 14:100.

(Prepared by D. H. Gent)

Black Root Rot

The oomycete *Phytophthora citricola* causes root rot and trunk canker on many economically important plants. Crown and root rot of hop ("black root rot") caused by *P. citricola* was first described in New Zealand, where it caused significant plant mortality following the move from more resistant English cultivars to the higher yielding, but susceptible, American Late Cluster during the 1930s. *P. citricola* was the main pathogen associated with hop root and crown rot in the Hallertau region of Germany. The disease has also been reported in South Africa, the United Kingdom, and the United States.

Symptoms

Infection usually begins at the crown and spreads to roots and bines. Within roots, infected tissue is watersoaked and blackened (Fig. 19) with a clear demarcation in color between infected and healthy tissue. Infection can spread from the crown for several centimeters up the base of the bine, with internal tissues appearing discolored when bines are split longitudinally. The vascular cylinder remains green. In the United States, black root rot has also been described as causing a thickening of the bases of bines. Severe infection of roots becomes apparent late in the season with a yellowing of foliage followed by sudden wilting of leaves, shoot tips, and lateral branches. Infection can cause wilting of some or all maturing bines. As the disease progresses, leaves turn black but remain attached to the bine and wilted bines remain firmly attached to the crown. Severely wilted plants often die over the winter. In Germany, the disease was characterized by a gradual rotting of the rhizomes of hop plants until only small parts of the crown remained healthy. Infection was not associated with wilting, but the main rhizomes and crown developed a dry rot that reduced bine growth, yield, and quality.

Plants affected by black root rot are often aggregated and associated with heavy soils or poorly drained areas within a field. Wet soil conditions encourage infection and disease progress. However, foliar symptoms may first become apparent under dry conditions when affected roots are unable to provide sufficient water for plants. Mechanical operations that wound roots and crowns are also conducive to infection.

Causal Organism

Black root rot is caused by the oomycete *P. citricola* Sawada, which lives in the soil and infects the underground portions of the hop plant. The main survival structures are globose sexual oospores, 16–30 µm in diameter, which can be produced in soil or in plant tissue. Chlamydospores are rare and not considered important for survival. Oospores may germinate directly or germinate to produce zoosporangia (30–75 × 21–44 µm), which are semipapillate, noncaducous, and variable in shape. Zoosporangia also are produced in sympodial fashion on simple sporangiophores borne on coenocytic mycelium Zoosporangia can germinate directly or produce flagellate zoospores (13 × 8–9 µm), which can survive for up to 24 h following release.

Disease Cycle and Epidemiology

The disease cycle of *P. citricola* on hop or many other hosts has not been studied in detail. *P. citricola* has limited saprophytic survival. Oospores can survive at least 18 months in field soil, with some studies indicating survival for up to 6 years. Oospores or zoosporangia germinate to infect the plant directly or the latter may release zoospores that move in water films and are chemotactically attracted to host roots. Zoospores encyst once they come in contact with the root and can survive in soil for up to 4 weeks prior to infection. Zoospores germinate and produce mycelium that infects the host tissue, producing more oospores or zoosporangia. Multiplex real-time polymerase chain reaction assays have been developed to distinguish *P. citricola* from several other species of *Phytophthora* in plant tissue, which will aid identification and further study.

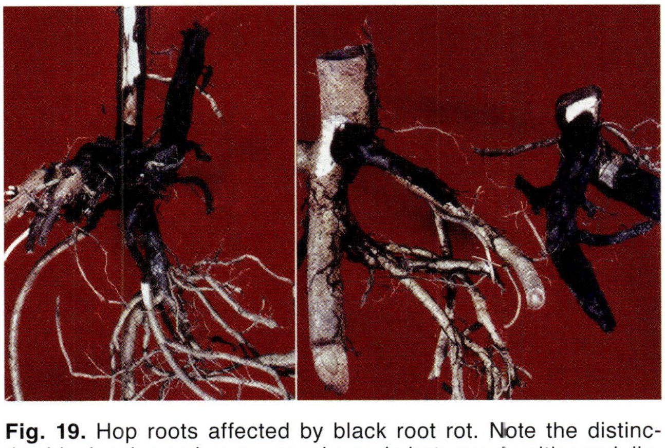

Fig. 19. Hop roots affected by black root rot. Note the distinctive black color and pronounced margin between healthy and diseased tissues. (Courtesy C. B. Skotland)

Management

In New Zealand, the adoption of the susceptible American Late Cluster cultivar in the 1930s led to increased plant mortality from black root rot. The disease was initially managed by rouging and replanting. However, this was not commercially satisfactory because American Late Cluster did not yield well in the first year after planting. By the 1950s, 20–60% of hills in yards with a high water table had to be replanted annually, until the release of resistant cultivars such as Smoothcone.

In the United States, cluster types E-2 and L-8 are considered extremely susceptible. L-1 and Galena are partially resistant and Alliance, Brewers Gold, Bullion, Calicross, Cascade, Columbia, Comet Eroica, Fuggle, Hallertauer, Northern Brewer, Nugget, Olympic, Talisman, Tettnanger, and Willamette highly resistant to black root rot in the field. In contrast, early reports in the United Kingdom described severe outbreaks of black root rot on Brewers Gold and infection of Fuggle and several Golding varieties, including Whitebread Golding Variety (W.G.V.). Reported differences in susceptibility of cultivars in different countries might indicate some differences in aggressiveness or pathogenicity of *P. citricola* strains. However, no studies have been conducted.

There are few reports of chemical control of *P. citricola* in hop. As with other diseases caused by *Phytophthora* spp., avoidance of waterlogging through good drainage and appropriate irrigation are important in reducing disease. Agronomic practices that cause wounding to crowns or roots are likely to increase infection by the pathogen. The adoption of reduced cultivation practices for weed control and control of hop growth in yards in recent years is likely to have provided some reduction in black root rot.

Selected References

Brien, R. M. 1938. Black root rot of hops. New Zeal. J. Sci. Technol. A20:62-64.

Erwin, D. C., and Ribeiro, O. K. 1996. Phytophthora Diseases Worldwide. American Phytopathological Society, St. Paul, MN.

Pichlmaier, J., and Zinkernagel, V. 1992. The black root and crown rot of hop and its causing agent *Phytophthora citricola* (Sawada). J. Plant Dis. Protect. 99:505-518.

Schena, L., Hughes, K. J. D., and Cooke, D. E. L. 2006. Detection and quantification of *Phytophthora ramorum*, *P. kernoviae*, *P. citricola* and *P. quercina* in symptomatic leaves by multiplex real-time PCR. Mol. Plant Pathol. 7:365-369.

Skotland, C. B. 1982. The reaction of hop, *Humulus lupulus*, varieties to *Phytophthora citricola*. Phytopathology 72:994-995.

Sonoda, R. M., and Ogawa, J. M. 1968. Phytophthora crown and root rot of hops in California. Plant Dis. Rep. 52:76-78.

(Prepared by F. S. Hay)

Cone Tip Blight

Cone tip blight was first reported in the Pacific Northwest region of the United States during the early 1990s on the cultivar Nugget, but it has since been found on cultivars Chinook and Willamette. Cone tip blight has also been reported in Australia on cultivars Agate, Nugget, and Willamette as well as in Poland on the cultivar Marynka. Incidence of affected cones is usually low (<1%), but surveys have detected higher incidence levels in Nugget (5–30%), Willamette (10–32%), and Agate (3–10%).

Symptoms

Affected bracts and bracteoles become necrotic as the cone approaches full maturity (Figs. 20 and 21), starting at the tip of the cone and moving up the strig. The extent of affected cone area varies; sometimes only the very tip of a cone may be necrotic, but generally all the bracts and bracteoles in a whorl tend to be affected. When the necrotic bracts and bracteoles are removed, necrosis of that portion of the strig is usually evident (Fig. 22).

Causal Organisms

Several *Fusarium* species may cause cone tip blight. In the United States and Poland, *F. avenaceum* (Fr.:Fr.) Sacc. (teleomorph: *Gibberella avenacea* R. J. Cook) and *F. sambucinum* Fuckel (teleomorph: *G. pulicaris* (Fr.:Fr.) Sacc.) are reported to cause cone tip blight. In Australia, the disease is caused by *F. crookwellense* L. W. Burgess, P. E. Nelson, & Toussoun.

Fig. 20. Brown discoloration of bracts and bracteoles characteristic of cone tip blight symptoms on cultivar Agate. (Courtesy S. J. Pethybridge)

Fig. 21. Discoloration of the tips of cones of cultivar Nugget affected by cone tip blight. (Courtesy J. C. Bienapfl)

Fig. 22. Discoloration of strigs from cones of cultivar Nugget affected by cone tip blight. (Courtesy J. C. Bienapfl)

F. avenaceum produces slender macroconidia with thin cell walls and an elongated apical cell that may be bent. Microconidia are rare and chlamydospores are absent. Dense aerial mycelium that is white or tan to reddish brown grows rapidly on potato dextrose agar. Sporodochia are orange when present. The undersurface of a colony on potato dextrose agar is carmine red or light to dark brown.

F. sambucinum produces short, stout, moderately to strongly curved macroconidia with thick cell walls. Each has a strongly curved, snout-like apical cell that often has a small papilla. Microconidia are rare, and abundant chlamydospores are quickly formed in chains, in clumps, or singly. On potato dextrose agar, mycelial growth is rapid, usually resulting in dense aerial mycelium that is pinkish to reddish brown, white, or tan. Sporodochia are creamy, tan, or orange when present. The undersurface of a colony on potato dextrose agar is commonly carmine red but may be light to dark brown.

F. crookwellense produces thick-walled macroconidia with a ventral surface less curved than the strongly arched dorsal surface. The basal cell is foot-shaped while the apical cell is curved with a narrow tip. Chlamydospores are formed in both hyphae and macroconidia; microconidia are absent. Dense, aerial mycelium that is white to tan grows rapidly on potato dextrose agar. Sporodochia produced are orange to red-brown. The undersurface of a colony on potato dextrose agar is carmine red.

Disease Cycle and Epidemiology

Little information is available on the disease cycle and epidemiology of cone tip blight. *F. sambucinum* survives in soil, plant debris, and affected hop crowns while *F. avenaceum* and *F. crookwellense* can survive in soil or plant debris. Field observations suggest that the disease may be more severe when high humidity conditions prevail during cone development. Spores may land on flowers during the burr stage even though disease symptoms are not evident until later in cone development, commonly just before harvest.

Management

Specific control measures have not been developed. Limited evaluations of fungicides have not been successful for management of cone tip blight. Measures that reduce humidity in the canopy may help to control the disease.

Selected References

Bienapfl, J. C. 2004. Hop cone tip blight: A new disease in the Pacific Northwest. M.S. thesis, Oregon State University, Corvallis.

Bienapfl, J. C., Ocamb, C. M., Klein, R., and Nelson, M. 2005. Fusarium cone tip blight of *Humulus lupulus*. Acta Hortic. 668:123-128.

Pethybridge, S. J., Hay, F. S., Wilson, C. R., Sherriff, L. J., and Leggett, G. W. 2001. First report of *Fusarium crookwellense* causing tip blight on cones of hop. Plant Dis. 85:1208.

Solarska, E. 2003. Control of *Fusarium* spp. on hop cones with grapefruit extract. Pages 32-35 in: Proc. Sci. Comm. Int. Hop Growers Comm., Dobrna-Žalec, Slovenia.

(Prepared by C. M. Ocamb and J. C. Bienapfl)

Downy Mildew

Downy mildew is a major disease in many hop-growing areas of the world. The disease was first reported on wild and cultivated hops on separate islands in Japan in 1905 and later in Wisconsin in 1909. It was found in the United Kingdom in 1920 and 7 years later had spread throughout the hop-growing areas of Europe. Downy mildew was first reported in New York and western Washington in 1928, in western Oregon in 1930, in California in 1934, and in the Yakima Valley of south-central

Washington in 1937. Currently, the disease has not been reported in Australia, New Zealand, or South Africa.

Yield and quality losses from downy mildew vary widely depending on cultivar susceptibility and weather conditions during vegetative growth and harvest. A complete loss in marketable yield is possible when significant cone infection occurs. Downy mildew was a factor in the decline and loss of hop production in New York, Wisconsin, and the coastal areas of California, and its occurrence in the high-rainfall areas of western Oregon and Washington led to the demise of the susceptible Cluster cultivars in those regions. In regions with high disease pressure, the hop industry has survived by growing more resistant cultivars and through extensive fungicide applications.

Symptoms

Downy mildew results in characteristic, systemically infected shoots termed "spikes." Spikes arising from the hop crown are called primary spikes or primary basal spikes and are stunted with shortened internodes and chlorotic to yellow-green leaves that are brittle and curl downward (Fig. 23). Symptoms on primary spikes occur on shoot tissue from the crown up. Diseased shoots arising from infected apical meristems are called secondary spikes. If they occur on lateral branches or the trained bines later in the season, they are called aerial spikes. Secondary spikes show symptoms similar to primary spikes, except plant tissues below the infected area remain normal in appearance (Fig. 24). Also, internodes may not be as shortened on secondary spikes as on primary spikes. Young, developing lateral branches that become infected are stunted and later become necrotic and desiccate during dry weather (Fig. 25). Infection of trained bines arrests their development, causing the bines to fall away from the string (Fig. 26). Localized lesions on abaxial surfaces of leaves are delimited by leaf veins and appear angular and watersoaked but quickly become necrotic and light to dark brown. Sporangia may form in a mass on the underside of infected individual leaves and spikes, forming a purple-gray to black downy growth (Figs. 27 and 28).

Infected inflorescences become dark brown, shrivel, dry up, and may fall. Cones become brown, harden, and cease further development when infected early in the season. Entire cones may become dark brown or only bracts may be discolored when infected later in the season, causing a distinctive striped appearance (Fig. 29). Masses of sporangia may or may not be present under diseased bracts and bracteoles.

Infected hop crowns exhibit reddish-brown to black flecks and streaks in the white crown tissue next to the bark (Fig. 30).

Fig. 23. Stunted and chlorotic hop shoots ("spikes") caused by systemic infection of *Pseudoperonospora humuli*. (Courtesy D. A. Johnson [left] and D. H. Gent)

Fig. 24. Secondary basal "spike" infected with *Pseudoperonospora humuli* showing chlorosis of leaves on the upper nodes of the affected shoot. (Courtesy D. H. Gent)

Fig. 25. Stunted lateral branch caused by systemic infection with *Pseudoperonospora humuli*, producing an aerial "spike." (Courtesy D. H. Gent)

This should not be confused with the reddish tissue found in the center of healthy hop crowns of some cultivars. Infected crowns of some susceptible cultivars may be in various stages of rot, ranging from almost completely destroyed to apparently healthy.

Causal Organism

Downy mildew is caused by *Pseudoperonospora humuli* (Miyabe & Takah.) G. W. Wilson, of the family Peronosporaceae and order Peronosporales. The host range of *P. humuli*

Fig. 26. Bines affected by downy mildew falling away from a string. (Courtesy D. H. Gent)

Fig. 27. Angular leaf lesions caused by *Pseudoperonospora humuli*. Adaxial, left, and abaxial, right. (Courtesy D. H. Gent)

Fig. 28. Leaf chlorosis associated with systemic infection by *Pseudoperonospora humuli*. (Courtesy D. H. Gent)

Fig. 29. Characteristic dark brown discoloration of cones caused by downy mildew. (Courtesy D. R. Smith)

is reported to include the cultivated hop (*Humulus lupulus*), *H. japonicus*, and certain nettles (*Urtica* spp.). Results of recent molecular analyses of *P. humuli* and *P. cubensis* (Berk. & M. A. Curtis) suggest that these organisms are closely related or synonymous; however, the organisms do not cross-infect cucurbits and hop. Polymerase chain reaction assays have been developed to aid in the identification of *P. humuli*.

Haustoria of *P. humuli* are branched and often appear stellate, particularly in young haustoria. Later in infection, host deposits of callose cause haustoria to appear knob-like. Sporangiophores vary from 190 to 430 µm in length and have determinate growth. The sporangiophores are dichotomously branched at acute angles and taper to curved, pointed tips on which sporangia are produced (Fig. 31). Lemon-shaped sporangia borne on the sporangiophores are 22–26 × 15–18 µm and germinate only indirectly by zoospores. Each sporangium discharges four to eight flagellated zoospores, each 7–10 µm in diameter. The sexually produced oospores can be found in infected soil, leaves, shoots, and particularly cones. Oospores are globular to spherical, have smooth walls, and are reported to range from 22 to 42 µm in diameter.

Disease Cycle and Epidemiology

The pathogen overwinters as intercellular mycelium in infected dormant buds and crowns (Fig. 32). Mycelium may spread into developing buds during the winter and early spring,

Fig. 30. Rhizomes with reddish-brown to black flecks and streaks associated with downy mildew. (Courtesy D. R. Smith)

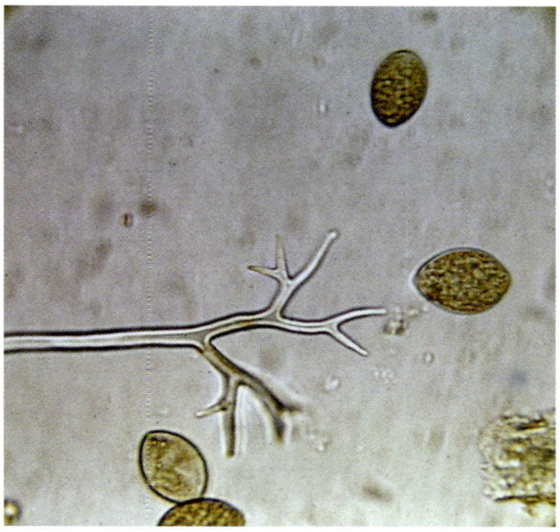

Fig. 31. Sporangiophore and sporangia of *Pseudoperonospora humuli.* (Courtesy D. R. Smith)

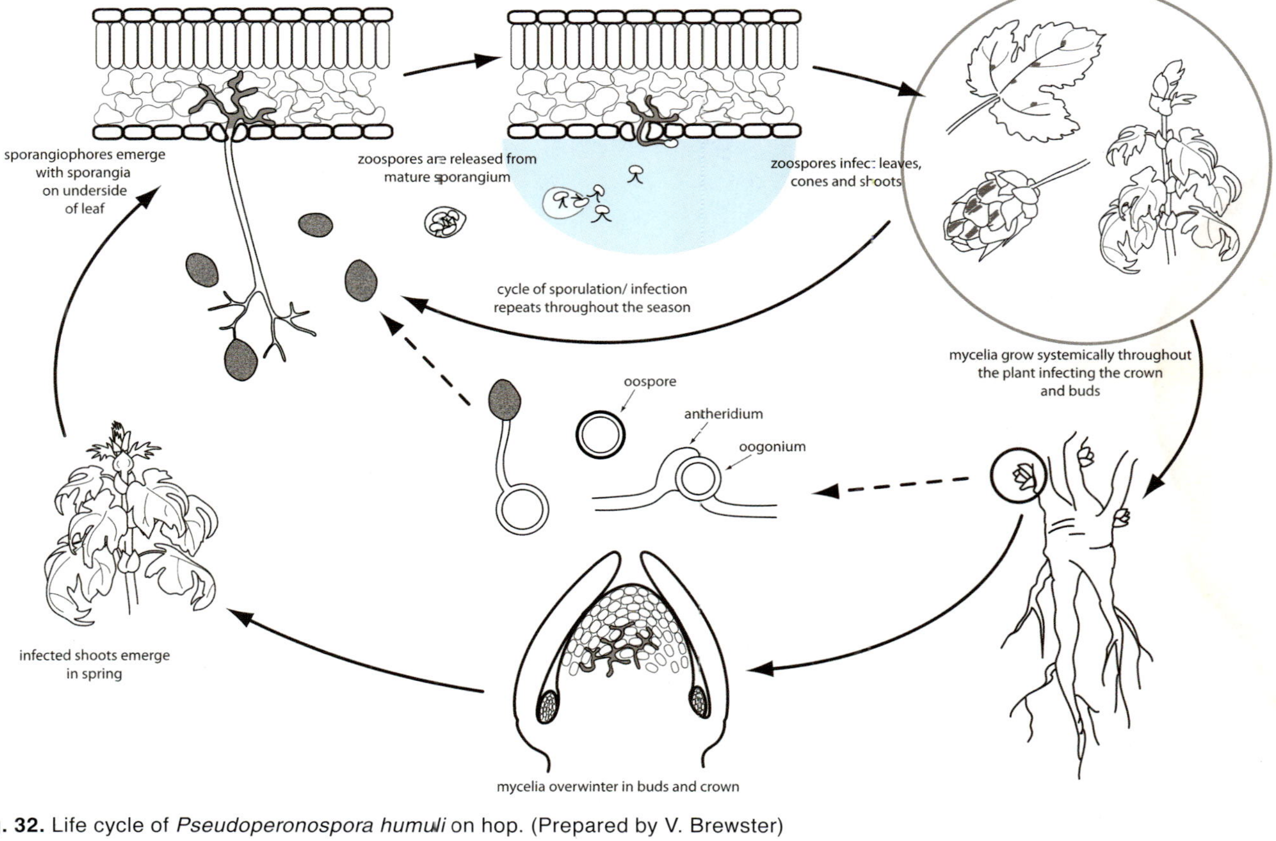

Fig. 32. Life cycle of *Pseudoperonospora humuli* on hop. (Prepared by V. Brewster)

which may result in the emergence of systemically infected shoots when plants break dormancy. These infected shoots are the primary basal spikes. Infected crowns may have any number of infected shoots or none at all. Sporangia are produced on the lower leaf surfaces of the primary spikes at night in response to a diurnal cycle when the temperature and humidity are favorable. Sporangia are released in mid-morning to early afternoon, especially during rainy weather, and germinate indirectly to produce zoospores. Zoospores enter hop tissues through open stomata and can infect leaves, bud stipules, apical meristems, and cones when the temperature is favorable and moisture is present. Infections of leaves result in localized leaf spots that can provide secondary inoculum for later infections of shoots, leaves, and cones, but leaf lesions tend to be ephemeral and desiccate quickly in dry weather. Infections of apical meristems become systemic, producing secondary spikes and additional sporaniga that may perpetuate the disease cycle. Mycelium progresses in the shoot tissue towards the crown during the growing season and can perennate in the crown and dormant buds. Carbohydrate reserves are reduced in systemically infected rhizomes, and weakened plants may die before harvest, during the winter, or after spring pruning operations.

The role of oospores in the disease cycle is not well understood. The importance of oospores as a source of primary inoculum has been largely discounted because attempts to germinate and infect plants have failed or results have been inconsistent. In New York, oospores were considered important in the overwintering of the pathogen. Oospores are found readily in infected hop tissue in most of the hop-growing areas of the world where downy mildew occurs. In arid climates, oospores are generally rare except in infected cones, where considerable numbers may be found.

Wet conditions from rain, dew, fog, or sprinkler irrigation and mild to moderate temperatures favor downy mildew, although dew wetness is thought to be of little importance in Europe. High humidity (>80–90%) and night temperatures greater than 5°C are required for sporulation; the optimal temperatures range from 16 to 20°C. Moisture on plant tissues is required for sporangium germination, zoospore motility and encystment, zoospore germination, and penetration of the germ tubes into the host tissue through stomata. Shoot infection requires a wet period of 3–6 h and proceeds within a temperature range of 8 to 23°C. Leaf infections require shorter durations of wetness than shoot infections, and leaf infections may occur within 1.5–2 h at 15–29°C. Leaf infections can occur at temperatures as low as 5°C when wet periods persist for 24 h or longer. However, wet periods need to persist at least 4–8 h at moderate temperatures for appreciable leaf and shoot infection to occur.

Management

Management of downy mildew requires the integration of strict sanitation practices, resistant cultivars, appropriate harvest timing, and timely applications of fungicides. Moisture on foliage favors downy mildew development, and practices that reduce periods of wetness help to minimize the disease. Overhead irrigation, especially during and after burr development, should be avoided.

Host resistance is the most efficient means of controlling downy mildew, but market forces limit the ability of growers to plant less susceptible cultivars. Cultivars vary in susceptibility to the crown rot phase of the disease and to leaf, shoot, and cone infections (Table 7). However, no cultivars are completely immune. Moderately resistant cultivars include Cascade, Fuggle, Perle, Tettnanger, and Willamette. Cultivars susceptible to the foliar phase of the disease include Cluster, Galena, Hallertauer Mittelfrüh, Hersbrucker Spält, and Nugget. However, susceptibility to crown rot varies among these cultivars, with Cluster being extremely susceptible.

Sanitation practices include the selection of noninfected rhizomes for planting, the removal of heavily diseased plants early in the season, and the elimination of primary basal spikes, mechanically or chemically, during spring pruning operations or manually during training. Pruning yards as late as possible generally reduces the severity of downy mildew, but pruning timing should be determined carefully to avoid reductions in yield from delayed training. Manual or chemical stripping of leaves after training reduces inoculum density and can minimize disease spread higher into the canopy. The success of sanitation practices depends on their thoroughness, and these practices help delay epidemics. An early harvest can also minimize cone infection when disease-conducive weather occurs late in the season, although yield in the current and subsequent seasons can be reduced with certain cultivars.

Timely fungicide applications can reduce the epidemic rate during favorable weather conditions. Fungicides should be applied prior to or soon after infection because once *P. humuli* has penetrated plant tissue it is less affected by nonsystemic fungicides. Several disease forecasting models have been developed in the United States and Europe to predict when downy mildew is likely to develop and can aid in scheduling fungicide applications. Resistant and moderately resistant cultivars require fewer fungicide applications than susceptible cultivars during weather that favors infection and disease development.

When managing downy mildew, tactics for fungicide-resistant management need to be employed. *P. humuli* produces numerous progeny during disease-conducive weather, and isolates resistant to a particular fungicide can develop and increase with repeated use of that fungicide. Isolates of *P. humuli* resistant to metalaxyl are widespread in areas of the United States and Europe, and metalaxyl should not be used where resistant populations are present. Reduced sensitivity to fosetyl-Al has been reported in the United States, and control of downy mildew requires higher rates of this compound where insensitive isolates are prevalent.

Selected References

Gent, D. H., Nelson, M. E., and Grove, G. G. 2008. Persistence of phenylamide insensitivity in *Pseudoperonospora humuli*. Plant Dis. 92:463-468.

Johnson, D. A., and Skotland, C. B. 1985. Effects of temperature and relative humidity on sporangium production of *Pseudoperonospora humuli* on hop. Phytopathology 75:127-129.

Johnson, D. A., Alldredge, J. R., Allen, J. R., and Allwine, R. 1991. Spatial pattern of downy mildew in hop yards during severe and mild disease epidemics. Phytopathology 81:1369-1374.

Johnson, D. A., Alldredge, J. R., and Allen, J. R. 1994. Weather and downy mildew epidemics of hop in Washington State. Phytopathology 84:524-527.

Kohlmann, H., and Kastner, A. 1975. Der Hopfen. Hopfen-Verlag, Wolnzach, Germany.

Kremheller, H. T., and Diercks, R. 1983. Epidemiologie und Prognose des Falschen Mehltaues (*Pseudoperonospora humuli*) an Hopfen. Z. Pflanzenkr. Pflanzenschutz 90:599-616.

Nelson, M. E., Eastwell, K. C., Grove, G. G., Barbour, J. D., Ocamb, C. M., and Alldredge, J. R. 2004. Sensitivity of *Pseudoperonospora humuli* (the causal agent of hop downy mildew) from Washington, Idaho, and Oregon to fosetyl-Al (Aliette). Online. Plant Health Progress doi:10.1094/PHP-2004-0811-01-RS.

Royel, D. J., and Kremheller, H. Th. 1981. Downy mildew of the hop. Pages 395-419 in: The Downy Mildews. D. M. Spencer, ed. Academic Press, New York.

Skotland, C. B. 1985. Infection of hop crowns and roots by *Pseudoperonospora humuli* and its relation to crown and root rot and overwintering of the pathogen. Phytopathology 51:241-244.

Skotland, C. B., and Johnson, D. A. 1983. Control of downy mildew of hops. Plant Dis. 67:1183-1185.

(Prepared by D. A. Johnson, B. Engelhard, and D. H. Gent)

Fusarium Canker

Fusarium canker was initially observed in English hop yards and subsequently described in 1902. This disease has also been referred to as "growing off," "fusariosis," "black-butt," and "wilt." Fusarium canker has been reported in commercial plantings across Europe, Australia, and the United States on many hop cultivars. Usually a low incidence of canker is found in commercial yards when this disease is present, but occasionally a high percentage of infected plants is observed within a yard.

Symptoms

Affected bines wilt rapidly and suddenly (Fig. 33). This wilt generally occurs during flowering or under high temperature conditions. Leaves may show chlorosis. The bases of affected bines are typically swollen from the disruption of photosynthate transport through the phloem. The swelling constricts off near the crown (Fig. 34) such that the point of bine attachment to the crown may consist of only a few central vascular elements. Affected bines are readily detached from the crown with a gentle tug. Occasionally, diseased bines also have a longitudinal split in the colonized cortical area (Fig. 35). Vascular discoloration is not associated with canker; usually a brown discoloration is found in affected cortical tissue of infected bines (Fig. 36). Cankers may also form on rhizomes of affected plants. An infected crown may completely die or result in stunted, unthrifty growth that eventually dies out.

Causal Organism

Fusarium canker is caused by *Fusarium sambucinum* Fuckel (teleomorph: *Gibberella pulicaris* (Fr.:Fr.) Sacc.). *F. sambucinum* produces short, stout, moderately to strongly curved macroconidia with thick cell walls and a strongly curved, snout-like apical cell that often has a small papilla. Microconidia are rare. Abundant chlamydospores are quickly formed in chains, in clumps, or singly. Mycelial growth is rapid on potato dextrose agar and usually includes dense aerial mycelium that is commonly pinkish, reddish brown, white, or tan. Sporodochia are creamy, tan, or orange when present. The undersurface of a colony on potato dextrose agar is commonly carmine red but may be light to dark brown. Perithecia of *G. pulicaris* are dark purple and ovoid to subglobose.

Disease Cycle and Epidemiology

Little information is available on the epidemiology of Fusarium canker. *F. sambucinum* survives in soil, plant debris, and diseased crowns and may be detected on apparently healthy plants propagated by softwood cutting as well as in soil with no history of hop production. It is thought that *F. sambucinum* infects hop plants primarily through wounds. Twisting and swaying of the bines caused by wind, tractors, and sprayers injure epidermal and cortical tissue at or below the soil line, and crowns can be wounded by cultivation or insect feeding, creating entry wounds for *F. sambucinum*. Under moist conditions, mycelia and sporodochia of *F. sambucinum* or perithecia of *G. pulicaris* may be observed on the outer surfaces of diseased bines, including remnants of affected bines from the previous season.

Fig. 33. Wilted bine with Fusarium canker and adjacent healthy bines. (Courtesy C. M. Ocamb)

Fig. 34. Cankered bines with swollen base and narrow point of attachment at crown. (Courtesy D. H. Gent)

Fig. 35. Longitudinal split associated with Fusarium canker. (Courtesy C. M. Ocamb)

Fig. 36. Left, cankered bine with swollen base. Right, bines with vascular browning associated with Verticillium wilt and bine from plant with Fusarium canker where vascular browning is absent. (Courtesy C. M. Ocamb)

The incidence of canker in hop yards is sporadic; not every bine on a hill may be diseased, nor are the same plants affected every year. Disease incidence may be greater where soils are wet or during growing seasons following wet winters.

Management

Removal of cankered tissue from affected hills and subsequent hilled soil around bines promotes the growth of healthy roots and can reduce the incidence of bine wilting. Cuttings should not be made from diseased hills, and diseased cuttings should be destroyed.

Practices that reduce moisture near the crown and injury to bines help to ameliorate Fusarium canker. Wounding healthy plants when planting should be avoided, and damage to bines from arthropod pests should be prevented. Minimizing injury to hops by restricting bine movement through training, arching, and row orientation can help reduce wounding and therefore pathogen entry into bines. Host resistance has been investigated, but no apparent resistance has been reported. In the United Kingdom, late spring application of a thiabendazole fungicide directed at basal shoots was reported to reduce the incidence of the disease.

Selected References

Darby, P. 1984. Plant Pathology. Rep. Dept. Hop Res. Wye Coll. 1983:22-27.

Davies, D. L. G. 1939. A rootstock canker of hops caused by *Gibberella pulicaris*. Annu. Rep. Long Ashton Res. Stn. Rep. 1938:115-223.

Royle, D. J. 1976. Plant pathology section. Rep. Dep. Hop Res. Wye Coll. 1975:25-29.

Rybáček, V. 1991. Hop Production. Vol. 16. V. Rybáček, ed. Elsevier Science, Oxford.

Sabo, J., Duric, T., and Jasnic, S. 2002. *Fusarium* fungi as a pathogen causing hop wilt. Plant Prot. Sci. 38:308-310.

Salmon, E. S., and Wormald, H. 1922. Hop "canker" or "growing off." J. S. E. Agric. Coll., Wye 24:354-359.

Schmidt, H. E., Naumann, K., and Holtz, S. 1969. Symptoms of hop wilting caused by *Fusarium sambucinum* Fuck. in the German Democratic Republic. Nachrichtenbl. Dtsch. Pflanzenschutzdienstes 23:121-126.

Wormald, H. 1946. Diseases of Fruits and Hops. 1st ed. Crosby Lockwood and Son, London.

(Prepared by C. M. Ocamb and J. C. Bienapfl)

Gray Mold

Gray mold is a problem mainly in growing regions with moist climates (e.g., Europe and Japan). The disease rarely occurs in regions with limited rainfall during the time from flowering to harvest (e.g., the Pacific Northwest in the United States). Gray mold can cause complete losses resulting from poor cone quality and subsequent crop rejection.

Symptoms

Symptoms (Fig. 37) are predominately associated with cones or burrs but can appear on leaves damaged by wind, hail, and insect feeding. Light brown spots form on cone tips and the outside stipular bracts and bracteoles and gradually spread during cone development. The bracteoles are more susceptible to damage than bracts, and diseased cones can develop a striped appearance. Symptoms are similar to those of Alternaria cone disorder but can be distinguished by the development of disease signs, specifically gray, cottony, sporulating mycelia on diseased tissue after long periods of wetness or high humidity (>90%). Signs can also become visible on stigmata that remain associated with cones or other damaged or necrotic plant tissue. Cones of infected plants are medium brown after drying.

Causal Organism

Botrytis cinerea Pers.:Fr. is the anamorph of *Botryotinia fuckeliana* (de Bary) Whetzel and is not specific to *Humulus* spp. It is a weak pathogen requiring an exogenous energy source (e.g., insect frass or senescent tissue) or a compromised host for successful infection. *B. cinerea* is a normal resident of and on necrotic, senescent, or dead tissue and can infect almost all dicotyledons, many monocotyledons, and some pteridophytes. The fungus grows on most media suitable for the general culture of fungi. *B. cinerea* is extremely variable genetically because it has multinucleate hyphae and spores and a heterothallic (both male and female) nature in some strains and because of sexual recombination. Hyphae are branched, septate, and hyaline, though they can appear tan when melanization occurs under moisture-limiting conditions. Long (generally 2–5 mm), dark brown conidiophores arise directly from hyphae, usually have a stout basal cell (16–30 μm), and are irregularly branched near the apex. Conidiophore growth is determinate with terminal cells swelling to form clavate ampullae from which conidia form synchronously as blastospores on very short denticles. Conidia are unicellular (8–14 × 6–9 μm), generally multinucleate, ovate, hyaline (appear grayish in mass), and smooth on the surface.

Sclerotia are considered the most important long-term survival structure, although this is debatable because some isolates of *B. cinerea* do not form sclerotia and survive saprophytically on senescent plant material. Sclerotia have been shown to survive for 25 weeks while conidia survived for 35 weeks in the field. Sclerotia are brown to black, irregularly shaped, and up to 5 mm in diameter. They can germinate to form apothecia

Fig. 37. Medium to reddish-brown discoloration of cones associated with gray mold. (Courtesy of B. Engelhard [left two images] and W. F. Mahaffee [right two images])

(sexual reproduction), conidiophores, or mycelia. Sclerotia most frequently germinate to form a succession of conidiophores for as long as 2 months at temperatures ranging from 3 to 27°C in the spring and fall.

Disease Cycle and Epidemiology

B. cinerea has a complex life cycle that can involve sexual and asexual reproduction. It has the capability to survive indefinitely as a saprophyte on organic material, to remain latent/quiescent in infected plants, and to sporulate at any point in its life cycle. The pathogen may also grow as an epiphyte. Dispersal is mainly through the dissemination of airborne conidia from sporulating mycelia, but it can also occur through ascospores, dispersion of infected debris, insects, and mechanical transmission. Conidia are released during rapid changes in relative humidity accompanied by air movement, splashing water, or mechanical disturbance. Once deposition has occurred, conidia can remain dormant for long periods of time or germinate if conditions are favorable. Germination requires a relative humidity >93% or free water and an exogenous nutrient source. It can occur at temperatures of 0–26°C; the optimal temperature is 20°C. Infection and disease development can occur at 0–35°C, but the optimal temperature is still 20°C. Cone damage, whether it is caused by mechanical injury, insect feeding, or other pathogens, increases the likelihood of infection. At any time during the infection process, *B. cinerea* can enter a state of quiescence where the pathogen no longer grows. The cause of quiescence is not understood, but it is thought to be associated with plant susceptibility or changes in environmental conditions.

Management

The main method for controlling gray mold is the application of fungicides. Applications should be made prior to long periods of leaf wetness. There are numerous products that have been shown to be effective against gray mold. However, fungicide resistance is common within *B. cinerea* populations, particularly to benzimidazoles and dicarboximides, and can develop quickly. Thus, strict resistance management practices need to be employed. A fungicide class should not be used more than twice in a row. As many different fungicide classes as possible should be used, and classes for which resistance is known to occur should be avoided. Cultural management of gray mold is best accomplished through practices that reduce the duration of cone wetness. Overhead irrigation should be avoided in the evenings. Row orientation and plant spacing at planting should be adjusted to increase airflow within the canopy without causing excessive mechanical damage to cones from wind or field operations. Insect feeding should also be reduced to help decrease the incidence of gray mold.

Selected References

Coley-Smith, J. R., Verhoeff, K., and Jarvis, W. R. 1980. The Biology of *Botrytis*. Academic Press, New York.

Jarvis, W. R. 1977. *Botryotinia* and *Botrytis* species: Taxonomy, physiology and pathogenicity. Can. Dep. Agric. Monogr. 15, Harrow, Ontario.

Johnson, K. B., Sawyer, T. L., and Powelson, M. L. 1994. Frequency of benzimidazole- and dicarboximide-resistant strains of *Botrytis cinerea* in western Oregon small fruit and snap bean plantings. Plant Dis. 78:572-577.

Sakai, M. 1976. Studies on grey mold of hop. [*Botrytis cinerea*]. II. Occurrence of grey mold and rainfall in flowering period. Bull. Brew. Sci. 22:1-7.

Yourman, L. F., and Jeffers, S. N. 1999. Resistance to benzimidazole and dicarboximide fungicides in greenhouse isolates of *Botrytis cinerea*. Plant Dis. 83:569-575.

(Prepared by W. F. Mahaffee and B. Engelhard)

Powdery Mildew

Powdery mildew, also previously called "white mold" or "red mold" and caused by *Podosphaera macularis*, is the oldest known fungal disease of cultivated hops. It occurs in most commercial production regions, although it does not occur currently in Australia, New Zealand, or South Africa, where strict quarantines are in place to prevent its introduction. Similar quarantine regulations failed in 1996 in the Pacific Northwest of the United States when powdery mildew was found in a greenhouse in Washington, and it later became endemic in all Idaho, Oregon, and Washington production regions.

There have also been reports of several other powdery mildew species (*Erysiphe cichoracearum* DC., *Golovinomyces cichoracearum* (DC.) V. P. Heluta, *Oidium* sp., and *Phyllactinia suffulta* (Rebent.) Sacc.) occurring on various *Humulus* spp. It is likely that the *Oidium* sp. is *Podosphaera macularis*, since the genus *Oidium* designates the asexual phase when the ascocarp is not present. It is not clear whether the reports of other species reflect true infections of *Humulus* spp. or whether these reports are the results of misidentifications or contaminating ascocarps being found on cones or leaves (but not actually infecting these tissues).

Symptoms

Powdery mildew can occur on all green tissues of the hop bine. The symptoms most commonly associated with the disease are powdery and glistening white colonies that are either discrete and circular on leaves or stems (Fig. 38) or nondiscrete, encompassing large areas of tissue or complete organs (Fig. 39). These powdery, white areas consist of hyphae, conidiophores, and conidia of the pathogen. Symptom expression appears to be dependent on the cultivar, weather conditions, and fertility. Young, rapidly expanding leaves can develop raised blisters (Fig. 40) 1–3 days prior to visible signs of infection on the upper leaf surface. Blister development appears to be associated with environmental conditions that favor succulent growth, such as long periods of low light levels, excessive fertility (particularly nitrogen), or high soil moisture. Infections that occur on the underside of the leaf under the same conditions often result in pale to chlorotic spots visible on the upper surface (Fig. 41). Some cultivars develop these symptoms regardless of environmental conditions. As colonies age, they develop a dull and granular appearance, and some cultivars later develop necrotic areas under the colony (Fig. 42). Initially, infected burrs and young cones have only the typical white powdery fungal structures visible and then develop brown necrotic tissue (Fig. 43). If cones are infected early in their development, they are usually distorted or reduced in size (Fig. 44), and chasmothecia

Fig. 38. Discrete powdery white colonies of *Podosphaera macularis.* (Courtesy W. F. Mahaffee)

(formerly termed cleistothecia) often develop on them prior to harvest (Fig. 45). If cones are infected late in development, visible symptoms are often not observed until near harvest or drying in the kiln (see Cone Early Maturity).

Causal Organism

The currently accepted taxonomic designation for the fungus that causes powdery mildew is *P. macularis* (Wallr.) U. Braun & S. Takam. Over the years, numerous designations have been used and are now considered synonyms. These include *Alphitomorpha macularis* Wallr.:Fr., *A. humuli* Wallr., *Desetangsia humuli* (DC.) Nieuwl., *Erysiphe alchemillae* Grev., *E. humuli* DC., *E. macularis* (Wallr.:Fr.) Fr., *Sphaerotheca alchemillae* (Grev.) L. Junell, *S. castagnei* Lév., and *S. macularis* (Ehrh.) Magnus.

The pathogen is an obligate parasite specific to *Humulus* spp. *P. macularis* does not infect strawberry, raspberry, blackberry, or numerous weeds that occur in hop yards. These plants were once thought to be infected by variants of *S. macularis*, now named *P. aphanis* (Wallr.) U. Braun & S. Takam. However, there is a possibility that isolates that infect *Humulus* spp. also infect *Cannabis* spp. (the only other members of Cannabaceae), in which case the reciprocal may also be true.

The fungus grows superficially, originating from ascospores or conidia, on the surface of hop tissues, penetrating only the epidermal cells to form haustoria. Conidia are ovoid-ellipsoid to doliiform (20–33 × 13–20 µm) and occur in chains on erect conidiophores 40–70 µm long followed by one to three shorter cells (Fig. 46). Chasmothecia are formed in the fall, when hyphae from two different mating types fuse, and appear as small brown to black specks on infected tissue. Chasmothecia formation occurs most abundantly on cones but also are found on leaf tissue and in crown buds. Ascocarps are globose and

Fig. 39. Severe infection of leaf and cone tissue by *Podosphaera macularis* with nearly complete coverage of tissues by white colonies of the fungus. (Courtesy D. H. Gent [top] and W. F. Mahaffee [bottom])

Fig. 40. Blistering on leaves of cultivar Willamette associated with infection by *Podosphaera macularis*. (Courtesy W. F. Mahaffee)

Fig. 41. Chlorotic areas on upper surface of leaves with corresponding *Podosphaera macularis* colonies underneath. (Courtesy W. F. Mahaffee)

Fig. 42. Old colonies of *Podosphaera macularis* with necrotic tissue underneath and dull appearance. (Courtesy W. F. Mahaffee)

50–120 µm in diameter, have a 10- to 20-µm thick cell wall, and contain a single ascus of four to eight spores and seldom two spores. Ascospores vary in size depending on the number in the ascocarp. They are generally 20–25 × 12–18 µm and can be up to 30 × 24 µm. The number of appendages radiating from the lower half of the ascocarp varies. These appendages are unusually long, ranging from 0.5 to 6 times and sometimes more than 9 times the diameter of the ascocarp, and they are generally straight, septate, and yellow to dark brown.

Several races of *P. macularis* infecting *Humulus* spp. have been identified depending on their ability to overcome specific resistance genes in the host (*Rb*, R_1–R_7). No single race has been identified that overcomes all known resistance genes. Race structure differs among hop-growing regions.

Disease Cycle and Epidemiology

P. macularis overwinters as chasmothecia or mycelia in buds (bud perennation). Chasmothecia are located mainly in soil and plant residue surrounding the hills, but they have also been shown to occur between bud scales. In the Northern Hemisphere, chasmothecia mature either before November in the year formed or in the following March/April. However, ascospore discharge has been shown to occur only after the beginning of spring in association with an average nighttime (9:00 pm–5:00 am) temperature >10°C with >1 mm of rain or with an average daily temperature >10°C and ≥2.5 mm of rain. It is not known what conditions are suitable for ascospore infection, but they are likely similar to those for other powdery mildews where average daily temperatures >10°C favor infection, and

Fig. 43. *Podosphaera macularis* infected burr (left) and young cones (center and right) with conspicuous white mycelia and conidia of the fungus. (Courtesy W. F. Mahaffee)

Fig. 44. Cones infected with *Podosphaera macularis*. Note white powdery appearance of mycelia and conidia on and/or under bracts and bracteoles. (Courtesy W. F. Mahaffee)

Fig. 45. Minute, brown chasmothecia intermixed with mycelia of *Podosphaera macularis* on leaves and cones. (Courtesy Doug Whitener [left] and W. F. Mahaffee [center and right])

Fig. 46. Glistening *Podosphaera macularis* conidiophores with conidia. (Courtesy W. F. Mahaffee)

leaf wetness is neither required nor detrimental. Initial signs of infection from ascospore release are usually individual colonies on leaves located <7 cm from the ground; most colonies are located on the undersides of leaves.

In regions or fields where bud perennation has occurred, signs of the disease are visible almost immediately after shoot emergence begins in the spring (Fig. 47). As a shoot emerges from an infected bud (often called a flag shoot), it rapidly becomes covered with sporulating masses of mycelia that cause the shoot to appear white (Fig. 48). The infected shoots are often stunted with distorted leaves and are rapidly overgrown by uninfected shoots in the hill (Fig. 49). Under these conditions, the disease spreads extensively within the hill, but there is little spread among hills because of the limited exposure of inoculum to wind (Fig. 50). Occasionally, warm spring temperatures result in shoot growth that is faster than fungal growth, and bud perennation results in the infection of only the bract tissue at each node (Fig. 51). When this type of growth occurs, there is usually rapid and extensive spread within and among hills prior to pruning and training.

After training and before bloom, disease tends to increase in the basal growth (<1 m from the soil) with some movement up the trained bines. However, in regions with significant winds and conditions favorable to disease, there can be extensive spread to the trained bines. Colonies on the trained bines tend to be located on the undersides of leaves, especially during consistently wet weather. There is extensive disease development throughout the growing season in the basal growth of hills that can serve as an inoculum source for other developing tissue. Rapid plant growth favors disease development, and once lateral branches begin to develop there is another increase in disease that continues, environment permitting, through flowering and cone development. The burr (flower) is particularly susceptible to infection, which may be more a function of inoculum deposition than physiological characters, since the burr has evolved to capture pollen grains similar in size (30 μm in diameter) to *P. macularis* conidia. As leaves age they become increasingly resistant to infections; however, no such ontogenic resistance develops in cones.

Disease development is enhanced by factors that increase succulent plant growth, such as low light caused by cloud cover or canopy density, excess fertility (particularly nitrogen), and high soil moisture. Disease is reduced by extended periods of leaf wetness, intense light, or periods of limited soil moisture.

Losses associated with powdery mildew are mainly a function of cone infections and cost of control, since even severe leaf infections do not appear to impact bine growth. However, the level of cone infection is directly correlated with the level of leaf infection. High disease levels during flowering can result in complete crop loss caused by burr and cone abortion and malformed cones. Since there are no alternative markets for any significant quantity of hops, relatively low disease levels can result in complete crop loss in aroma cultivars if cone infections result in poor cone quality (browning and poor aroma) with subsequent rejection by buyers.

Asexually produced conidia that are carried by the wind are the primary dispersal agents. Rain splash and mechanical transfer (insect, tractor, or human activity) can also disseminate the pathogen at most life stages.

The primary environmental variable impacting disease development appears to be temperature. Infection is possible at constant temperatures of 8–28°C, while optimal infection, growth, and sporulation occur at 18–21°C. Each day, 3×10^5 conidia/cm^2 of colony are produced, and the potential exists for up to 40 generations during a growing season. The latent periods are 5, 10, 17, and 21 days at 18–28, 12–15, 10, and 8°C, respectively. There is an inverse relationship between infection, growth, or sporulation and temperature when it is >24°C. Infection and sporulation are reduced by 50% when temperatures exceed 30°C for periods of more than 3 h during the day, and young colonies (≤48 h old) may be killed by temperatures >36°C for 6 h during the day. The duration of leaf susceptibility is also reduced when temperatures are >30°C for 6 h or more during the day.

On susceptible cultivars under optimal conditions, a germ tube with appressorium (in this case a subtle thickening of the germ tube) emerges from a conidium within 6 h of deposi-

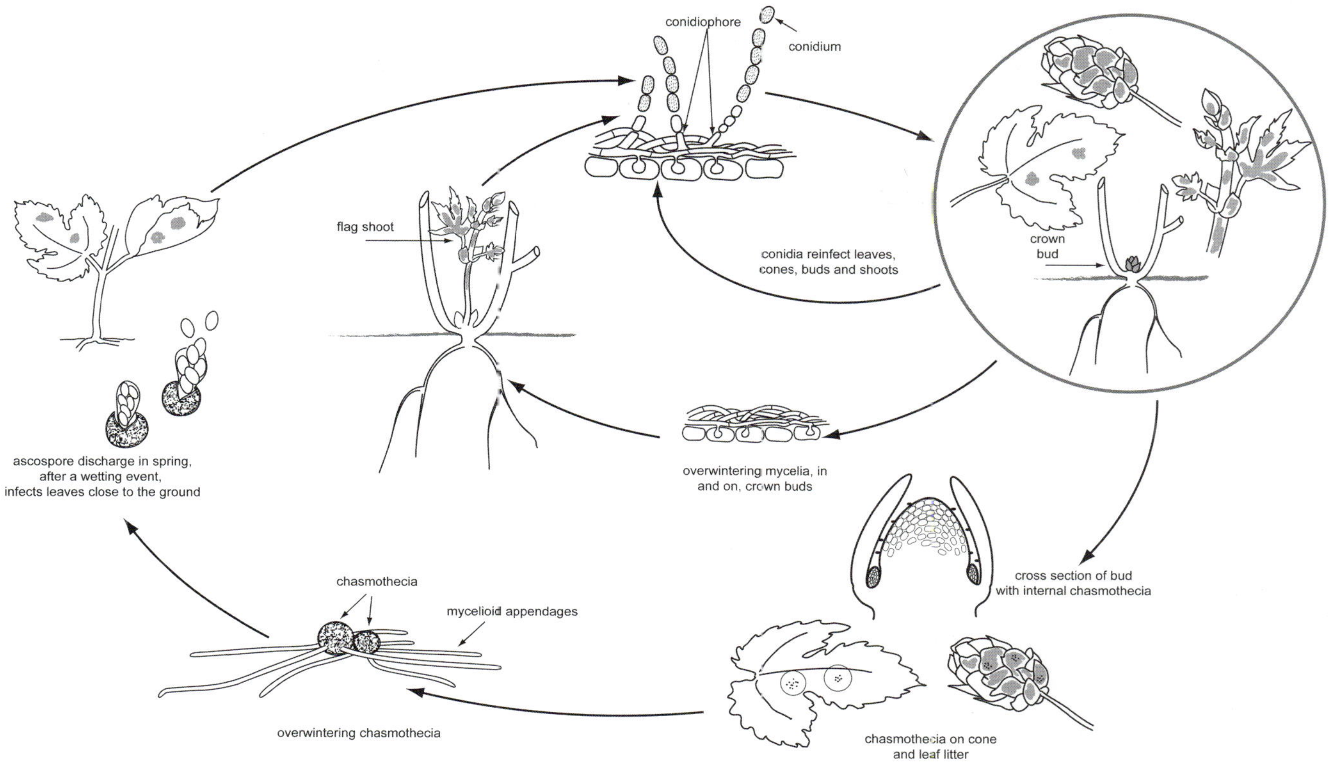

Fig. 47. Life cycle of *Podosphaera macularis* on hop. (Prepared by V. Brewster).

Fig. 48. Hop shoots severely infected with *Podosphaera macularis*, commonly termed flag shoots. (Courtesy W. F. Mahaffee, [left and center] and D. H. Gent [right])

Fig. 49. Newly emerged hop shoots infected with *Podosphaera macularis*. (Courtesy W. F. Mahaffee)

tion on the leaf surface, and a haustorium is present within 12–15 h. After 48 h, two to five germ tubes emerge and hyphal branching occurs. By the time 5 days have passed, mature conidia are available for dissemination, and sporulation intensifies as the colony continues to expand for 14–21 days. The duration of sporulation and colony expansion is dependent on environmental conditions and the age of the leaf at the time of infection. Rapidly expanding leaves tend to have larger colonies.

On most resistant cultivars, germ tube formation is generally delayed, appearing within 8–10 h, and the germ tube either fails to penetrate the epidermal cell or forms an irregular haustorium and fails to grow any further. On some cultivars, all steps of infection are delayed, mycelial growth and sporulation are retarded, and epidermal cells collapse, becoming chlorotic

Fig. 50. A hop plant severely infected with *Podosphaera macularis* prior to training (left, undisturbed and right, after two top leaves removed). (Courtesy W. F. Mahaffee)

Fig. 51. Hop shoots with *Podosphaera macularis* infection only occurring on the tissue surrounding each node. (Courtesy D. H. Gent)

and later necrotic areas. In some cases, tissue collapse occurs in advance of hyphae and the colony dies.

Management

Control of powdery mildew in commercial production is primarily accomplished through an intensive preventative fungicide program supplemented with cultural practices. Successful control is dependent on early and continuous intervention throughout the growing season. Fungicides can be applied according to a calendar program or by using an infection risk forecasting model (HOPS) developed for the Pacific Northwest hop-growing regions in the United States. Fungicides with activity against this pathogen include boscalid, fenarimol, myclobutanil, pyraclostrobin, quinoxyfen, spiroxamine, sulfur, tebuconazole, and trifloxystrobin. Other products such as various petroleum and plant oils, soaps, and sodium bicarbonate also have direct activity against *P. macularis*, while applications of copper and phosphite salts appear to reduce plant susceptibility to infection. Resistance management should be practiced by keeping disease levels as low as feasible, by diversifying the fungicide class (mode of action) used, by using each class only two to three times in a season, and by tank-mixing with fungicides with high risk for resistance development with those fungicides having a low risk.

Since the burr is highly susceptible to mechanical and other damage during pesticide applications, attempts should be made to minimize applications during this time. To reduce the chance of disease development during burr development, basal growth should be removed just prior to flowering and a highly efficacious protectant fungicide with a long residual action should be applied.

Cultural control includes sufficient but not excessive fertility and water management, reduction of initial inoculum, and removal of basal growth mechanically or chemically. Resistant cultivars have been developed but are often not acceptable because of market factors. Management of initial inoculum depends on whether overwintering occurs as chasmothecia or bud perennation. For regions where chasmothecia develop, initial inoculum may be reduced by removing diseased tissue from the field, through cultivation, or by applying compost to bury chasmothecia after the harvest. In regions where bud perennation occurs, removing buds 5–7 cm below the soil line prior to shoot emergence (crowning) significantly reduces the number of flag shoots in the spring. Alternatively, if spring pruning completely removes green tissue, disease development is reduced nearly as well as through crowning. Regardless of the method used for managing initial inoculum, banding an efficacious fungicide onto the hills before hill closure (the point at which the ground is not visible through the shoots) helps to reduce disease spread within the hill prior to training or pruning.

Because highly susceptible tissue and conditions favorable for disease development are abundant, periodic removal (mechanical or chemical defoliation) of basal growth is recommended once bines have grown far enough up the string (2.5–3 m) so that the growing tip will not be damaged. It is very important that, from training onward, any disturbance of the basal growth be immediately preceded by a protective fungicide application to reduce the chance that inoculum in the basal growth will spread to foliage on the trained bines. Applying excessive amounts of nitrogen or water should always be avoided since they may cause a flush of succulent growth, which is highly susceptible to infection.

When powdery mildew is present on cones near harvest, a timely harvest minimizes losses in cone quality caused by cone early maturity. An early harvest can also help to reduce damage to cones, but it may reduce yield in the current and following season on some cultivars.

Selected References

Liyanage, A. d. S., and Royle, D. J. 1976. Overwintering of *Sphaerotheca humuli*, the cause of hop powdery mildew. Ann. Appl. Biol. 83:381-394.

Mahaffee, W. F., Turechek, W. W., and Ocamb, C. M. 2003. Effect of variable temperature on infection severity of *Podosphaera macularis* on hops. Phytopathology 93:1587-1592.

Royle, D. J. 1978. Powdery mildew of the hop. Pages 381-409 in: The Powdery Mildews. D. M. Spencer, ed. Academic Press, London.

Seigner, E., Seefelder, S., Haugg, B., Engelhard, B., Hasyn, S., and Felsenstein, F. G. 2003. Potential of powdery mildew (*Sphaerotheca humuli*) to infect hops (*Humulus lupulus*) in various developmental stages. Ges. Pflanz. 55:29-33.

Turechek, W. W., and Mahaffee, W. F. 2004. Spatial pattern analysis of hop powdery mildew in the Pacific Northwest: Implications for sampling. Phytopathology 94:1116-1128.

(Prepared by W. F. Mahaffee, B. Engelhard, D. H. Gent, and G. G. Grove)

Red Crown Rot

Red crown rot (also known as red root rot) was first documented in Victoria, Australia, in 1981 and was confirmed in the United States in 2007. The disease is a dry rot leading to the deterioration of the crown and root system and may reduce yields by as much as 50%. The productive life of diseased plants is reduced from 30 to 15 years. Extensive replanting was necessary to reduce yield loss, but the benefit was only short term before the new plants were also infected. In Australia, symptoms first appeared in 2-year-old plants, where stem rot extended into roots and crowns. In 3-year-old plants, up to 30% of the roots and crown were destroyed. Loss of vigor from this disease was generally not observed until plants were at least 5 years old. In the United States, reduced vigor and erratic growth has been observed during establishment of hop yards when rhizomes were infected with the red crown rot pathogen.

Fig. 52. Hop roots affected by red crown rot displaying a dark reddish-brown discoloration of the pith tissues. (Courtesy P. McGee [top and bottom left] and D. H. Gent [bottom right])

Symptoms

Roots and crowns of diseased plants are gnarled and twisted and covered in loose, thick, brown bark. The affected core and larger roots are orange to red, and the tissue becomes dry and friable (Fig. 52), crumbling easily when held. The margin between the diseased and healthy tissue is generally well defined, although it may appear watersoaked and may also have a pink-red stain in the healthy tissue bordering the rotted tissue. Dark black fruiting bodies of the pathogen (pycnidia) are sometimes observed in diseased tissues.

Aboveground symptoms include uneven, thin, weak bines (Fig. 53) and chlorotic leaves. Leaves may also have a characteristic "parrot beak" leaf curl. Cone yields can be reduced by 20% or more, and alpha-acids are reduced from 10 to 8%.

Causal Organism

Red crown rot is caused by a *Phacidiopycnis* sp. (Fig. 54). The fungus produces hyaline, guttulate, spindle-shaped conidia (8–12 × 4–6 μm) within white cirrhi. On potato dextrose agar, mycelium is initially white, but it quickly becomes mouse gray. Black, irregularly shaped sclerotia form readily in culture.

Disease Cycle and Epidemiology

Farm procedures in hop yards in Victoria are believed to reduce plant vigor and contribute to disease severity. Compared with Tasmania, the growing season in Victoria is shorter and

Fig. 53. Reduced growth and death of hop plants caused by red crown rot. (Courtesy D. H. Gent)

Fig. 54. Fruiting bodies and mycelium associated with red crown rot caused by a *Phacidiopycnis* sp. (Courtesy P. McGee)

harvests occur earlier, which increases stress on the plants. Injury to the crown (e.g., during spring mowing of shoots and while anchoring support strings through the crown) are thought to be necessary to produce disease symptoms.

The causal organism is believed to live in and on soil, plant debris, and healthy crowns. Although symptoms can be found in the crowns and roots, the frequency of isolation of *Phacidiopycnis* sp. is considerably higher from diseased crowns, and this is the only pathogen found at the margin of diseased tissue.

Management

Fungicides are effective in reducing disease intensity under in vitro conditions but have failed to provide effective control in the field. While a cure for this disease has not been established, a range of management strategies can be used to minimize yield losses. Growers are advised to source planting material from areas certified free of the disease and rogue diseased plants and to avoid replanting in the hole left by removal of a diseased plant. Delaying the harvest by up to 14 days can increase yields by nearly 20% in affected hop plants and produces larger, more vigorous plants. Cutting bines above 1.5 m at harvest provides marginal but consistent increases in yield. Avoiding injury to the crown and roots during spring mowing and pruning operations can increase yield by approximately 10% when red root rot is present. Fumigation at planting can provide an increase of approximately 60% in yields in year one and 14% in the second year.

Selected Reference

McGee, P. A. 1997. Epidemiology of hop red root rot and strategies for control. Ph.D. diss. Royal Melbourne Institute of Technology, Melbourne, Australia.

(Prepared by P. McGee, A. Lawrie, and I. Porter)

Septoria Leaf Spot

Septoria leaf spot has been reported on hop in Kashmir (India) and in New Zealand. It can be damaging to hop in tropical or subtropical regions of production but is not prevalent in most counties that produce hops. The limited hop production in Kashmir has virtually ceased because of political unrest, and the disease has not been reported elsewhere in the past 20 years.

Symptoms

Disease symptoms appear initially as ash gray, circular to irregularly shaped lesions 1–3 mm in diameter with dark brown margins on older and lower leaves. Minute black specks (pycnidia) are apparent in leaf spots when viewed under 10× magnification. Leaf spots later become more numerous and appear on younger leaves higher on the plant. As leaf spots coalesce, large necrotic lesions may form and cause premature defoliation of leaves. Disease symptoms are reported to be more prominent on female plants.

Causal Organism

Septoria leaf spot is caused by *Septoria humuli* Westend. Pycnidia are dark and spherical, have a diameter of 82–114 μm, and have a prominent ostiole. Conidia are slender, filiform, hyaline, septate, and 37–53 × 1.3–2.5 μm.

Disease Cycle and Epidemiology

The disease cycle and epidemiology of Septoria leaf spot have not been investigated. In India, disease symptoms appear during April to June and coincide with moderate to warm (20–25°C) temperatures and moist weather. Presumably, the pathogen overwinters on crop debris and is dispersed by splashing water and wind, as with other *Septoria* spp. on pathogens from *Cannabis sativa* L. and other hosts.

Management

Control of Septoria leaf spot is rarely necessary in most regions of hop production. Prophylactic applications of broad-spectrum fungicides are reported to control the disease in India. Certain broad-spectrum fungicides applied for control of downy and powdery mildew are probably effective against Septoria leaf spot.

Selected References

Munjal, R. L., and Gautam, S. R. 1979. Laboratory evaluation of fungicides in controlling Septoria leaf spot of hop. Indian J. Mycol. Plant Pathol. 8:204.

Pennycook, S. R., and Galloway, D. J. 2004. Checklist of New Zealand Fungi. Fungal Diversity Press, Hong Kong.

Puttoo, B. L., Koul, M. L, and Saraf, R. L. 1975. Septoria leaf disease of hops (*Humulus lupulus*) in Kashmir. Indian Phytopathol. 27:247-248.

(Prepared by D. H. Gent)

Sclerotinia Wilt

Sclerotinia wilt, also referred to as white mold or Sclerotinia canker, affects nearly 400 weed and crop plant species, mostly herbaceous dicots. It is an occasional problem on hop but can cause severe damage during certain years. Cultivars Fuggle and Bramling appear especially susceptible. Although the disease occurs on some hosts worldwide, reports of Sclerotinia wilt on hop are limited to Europe and New Zealand.

Symptoms

Disease symptoms generally appear in late spring or early summer as soft, watersoaked lesions on bines just below or near the soil surface at the crown. Occasionally, lesions may be observed on bines 1–2 m above the crown. The tissue underlying the lesion collapses, creating a light brown to grayish lesion or canker 2.5–10 cm in length. Fluffy, white mycelia may be prominent during humid weather or on lesions near the soil surface (Fig. 55). Hardened, black sclerotia form on affected bines

Fig. 55. Bleached lesion with fluffy white mycelia (left) and sclerotia (right) of *Sclerotinia sclerotiorum* at the base of a hop bine. (Courtesy J. T. Smith)

and are diagnostic (Fig. 55). As the disease progresses, lesions expand and eventually girdle the entire bine and cause the plant to wilt. Leaves generally remain green until the bine is girdled completely. Sclerotia form abundantly in the pith cavity of affected bines near lesions. Disease symptoms may appear similar to those caused by Fusarium canker, but unlike Fusarium canker, mycelia and sclerotia are abundant in the pith cavities of bines affected by Sclerotinia wilt.

Causal Organism

Sclerotinia wilt is caused by *Sclerotinia sclerotiorum* (Lib.) de Bary. Sclerotia have a dull to shiny black exterior and a white to beige interior, are oblong to irregularly shaped, and are 5–25 × 2–7 mm. Apothecia develop from conditioned sclerotia in or on the soil. They appear as brown cylindrical stalks up to 25 mm in length that support a pale orange to tan dish- or funnel-shaped cap 2–9 mm in diameter. Numerous cylindrical asci containing eight ascospores are borne in each apothecium. An apothecium may release more than 8 million ascospores, and one or more apothecia may emerge from a single sclerotium. Ascospores are aseptate, elliptical, hyaline, and usually biguttulate.

Disease Cycle and Epidemiology

The disease cycle and epidemiology of Sclerotinia wilt on hop have not been investigated thoroughly but are presumably similar to those of the pathogen in other agricultural systems. The pathogen overwinters as sclerotia in infested crop debris and in the soil. Conditioned sclerotia within 5 cm of the soil surface germinate and produce apothecia during cool to moderate weather (15–25°C) when the soil surface remains moist for several days or longer. Generally, this occurs when abundant, lush plant growth is present to shade the soil. Ascospores ejected forcibly from apothecia are dispersed by wind or splashing water and colonize senescing or healthy plant tissues near the crown. At least 48 h of wetness is required for germination of ascospores. Sclerotia may also germinate and infect directly. Severe epidemics of Sclerotinia wilt on hop are reportedly associated with the hilling of soil contaminated with sclerotia onto crowns and with frost injury of developing basal buds, which apparently creates an infection court for the pathogen. New sclerotia formed in and on infected bines are returned to the soil, where they may persist in excess of 5 years. The pathogen may survive on numerous broadleaf weeds in and around hop yards.

Management

Control measures for Sclerotinia wilt of hop are usually not needed. In yards with a history of the disease, cultural practices that reduce the duration of wetness on plants and the soil surface, such as the removal of excess basal shoots, stripping of bines, and appropriate timing of irrigation, provide some reduction in the disease.

Selected References

Burgess, A. H. 1964. Hops: Botany, Cultivation, and Utilization. Leonard Hill, London.

Salmon, E. S., and Ware, W. M. 1936. Sclerotinia 'canker' of the hop. J. S. E. Agric. Coll. Wye 37:25-26.

(Prepared by D. H. Gent)

Sooty Mold

Sooty mold may affect hop and numerous other plants wherever they are grown. Sooty mold generally causes little damage to hop plants directly, but a low incidence of sooty mold in cones reduces crop quality and marketability. Severe sooty mold contamination reduces the photosynthetic area of the plant and may result in the loss of the entire crop because affected cones are rendered unmarketable.

Symptoms

Before the appearance of sooty mold, plant tissues become covered with a sticky honeydew substrate and develop a shiny appearance, which is caused by colonization and feeding by aphids. Signs and symptoms of sooty mold soon develop on this honeydew as a flattened, black mass of melanized fungal hyphae that resembles a fine layer of soot. Burrs and developing cones may later become covered with honeydew and quickly become black and sooty in appearance. Entire bracts, bracteoles, and lupulin glands may become black and sticky, but sooty mold tends to be most prevalent on the undersides of bracts and bracteoles and on plant tissues shaded from the sun. Damson-hop aphids and their castings may be evident (see Fig. 108).

Causal Organisms

Sooty mold is caused by a complex of fungi that have not been fully described on hop. The problem is attributed most often to various *Cladosporium* spp. including *C. fumago* Link and *C. herbarum* (Pers.:Fr.) Link. However, *Aureobasidium* and *Fumago* spp. have been implicated in sooty molds on other hosts. Many *Cladosporium* and *Aureobasidium* spp. are cosmopolitan saprophytes, are found commonly on the leaf surface of hop and numerous other plants, and are prevalent in the air of hop yards. These fungi are tolerant to a broad range of environmental conditions and stresses.

Disease Cycle and Epidemiology

The appearance of sooty mold is linked inseparably to the presence and development of the aphids, most importantly the damson-hop aphid. The insect probes the phloem strands of hop plants and sucks copious volumes of plant fluids by means of its long stylet. Damson-hop aphids consume more plant fluids than can be processed by their digestive systems, and they expel a dilute solution of excess carbohydrates, amino acids, and other substances (collectively referred to as honeydew) that may cover all plant parts, as well as other surfaces nearby. Honeydew provides an excellent medium to support the growth of numerous dark-pigmented fungi that grow superficially, but conspicuously, on all plant parts and nearby surfaces. The sooty mold fungi survive saprophytically on crop debris and decaying organic matter.

Management

Control of aphids controls sooty mold successfully. Resistance to certain insecticides is widespread in the damson-hop aphid.

Selected Reference

Neve, R. A. 1991. Hops. Chapman and Hall, London.

(Prepared by D. H. Gent)

Verticillium Wilt

Verticillium wilt is an important vascular disease of many herbaceous and woody plants throughout much of the temperate and subtropical regions. A mild (fluctuating) form of wilt of hops, caused by *Verticillium albo-atrum*, was first described in the United Kingdom in 1924. To date, mild Verticillium wilt caused either by *V. albo-atrum* or *V. dahliae* has been reported on hop in Germany, Poland, Belgium, France, New Zealand, Australia (Tasmania), and the United States (Oregon and Washington). In 1933, more severe symptoms and plant death were reported in the United Kingdom. These severe symptoms became known as lethal (progressive) wilt and caused significant economic damage to the English hop industry. In 1997, the first report of lethal wilt outside the United Kingdom came from Slovenia.

Symptoms

The appearance of the mild or lethal forms of the disease depends on an interaction of pathogen virulence, cultivar susceptibility, and environmental factors. Verticillium wilt caused by *V. dahliae* is relatively rare and causes only mild symptoms. *V. albo-atrum* is responsible for the majority of both forms of Verticillium wilt outbreaks. Symptoms common to both forms of Verticillium wilt include yellowing and wilting of the leaves, which starts at the bine base and advances upwards. Initially, the interveinal area between the main veins becomes chlorotic; it then progresses to marginal and interveinal necrosis (Fig. 56). Leaf margins turn upward, and the symptomatic leaves fall easily from the plant. Bines with symptomatic leaves tend to have a rough epidermis with swelling up to 1.5 m from the base and brown discoloration of the vascular tissue. However, the symptomatic bines do not develop a restriction near the soil as with Fusarium canker.

Fig. 56. Yellowing and wilting of leaves caused by Verticillium wilt. (Courtesy S. Radišek)

Mild wilt develops in susceptible hop cultivars infected by less virulent strains of the pathogen or in resistant cultivars infected by highly virulent strains. The intensity of symptoms varies from season to season such that plants symptomatic in one season may look healthy the next season. Symptomatic plants are usually scattered throughout a hop yard in areas with excessive soil moisture. The first leaf symptoms typically appear after flowering, and only some of the bines from a hill are symptomatic (Fig. 57). Leaves from lateral shoots arising from the axial buds of affected leaves are frequently asymptomatic. There is usually considerable thickening of the bines (Fig. 57) and browning of the central vascular tissue (Fig. 58). Plant death is rare.

Fig. 57. Mild form of Verticillium wilt. Note swelling of affected bine and healthy bine intertwined. (Courtesy S. Radišek)

Lethal wilt is associated with sensitive hop cultivars infected by highly virulent strains. Plants that are infected during the season usually develop symptoms during cone development. Within 2–3 weeks, there is extensive and rapid dieback of leaves and lateral shoots. The leaves usually fall while the cones remain attached and wither with the bine (Fig. 59) as the entire plant dies. Early in the disease development, the entire vascular tissue of the main bines develops a brown discoloration (Fig. 60). Occasionally, symptomatic plants survive through the winter to produce a few weak bines that soon develop the typical symptoms (Fig. 61) and die. Disease incidence and spread in hop yards is related more to the cultivation pattern than to environmental factors.

Causal Organism

V. albo-atrum Reinke & Berthier and *V. dahliae* Kleb produce abundant hyaline conidiophores that are verticillately

Fig. 58. Transverse and longitudinal sections of bine infected with Verticillium wilt. Brown vascular tissue is limited to the center, which is typical of the mild form of the disease. (Courtesy S. Radišek)

branched with two to four phialides arising from each node (Fig. 62). *V. albo-atrum* on natural host material forms conidiophores up to 300 µm tall that have dark, swollen bases. This differs from *V. dahliae*, whose conidiophores are shorter (80–160 µm) and have a hyaline base. Phialides vary in size and are mainly 20–30 × 1.4–3.2 µm in *V. albo-atrum* and 16–35 × 1–2.5 µm in *V. dahliae*. Conidia arise singly but congregate in small droplets at the apices of the phialides. Conidia are hyaline, ellipsoid to irregular, subcylindrical, single celled, rarely one septate, 3.5–10.5 × 2–4 µm in *V. albo-atrum*, and 2.5–8 × 1.4–3.2 µm in *V. dahliae*. The production of characteristic resting structures allows these two species to be distinguished. *V. albo-atrum* is identified by its dark resting mycelium (Fig. 63) formed from hyphae that swell, undergo septation, and develop melanized cell walls. Microsclerotia and chlamydospores are absent. In contrast, *V. dahliae* is able to form dark microsclerotia (Fig. 64) that arise from single hyphae by repeated budding and consist of swollen, almost globular cells. Microsclerotia vary from elongate to irregularly spherical and are 15–50 µm in diameter. Dark resting mycelia and chlamydospores are absent.

Fig. 59. Lethal form of Verticillium wilt. Leaves are fallen while cones stay and wither with the bine. (Courtesy S. Radišek)

Fig. 60. Longitudinal section of bine infected with Verticillium wilt (left) and healthy bine (right). Note brown discoloration of the entire vascular tissue, which is typical of the lethal form of the disease. (Courtesy S. Radišek)

Both pathogens can be isolated readily from the basal segments of symptomatic bines by placing pieces of vascular tissue onto a general fungal culture medium (e.g., potato dextrose agar or prune lactose yeast agar).

The emergence of highly virulent *V. albo-atrum* hop isolates is well documented in the United Kingdom. One mild (M) and three lethal (PV1, PV2, and PV3) *V. albo-atrum* isolates have been identified by using virulence testing on differential sets of hop cultivars. The occurrence of such virulent strains of *V. albo-atrum* has been reported from Slovenia, where two pathotypes have been identified and classified as PG1, which causes mild symptoms, and PG2, which causes lethal wilt on susceptible hop cultivars.

Disease Cycle and Epidemiology

The life cycle of *V. albo-atrum* and *V. dahliae* can be divided into dormant, parasitic, and saprophytic stages. Dormant resting mycelia and microsclerotia are stimulated to germinate in soil, probably by root exudates of both host and nonhost plants. The fungi enter the roots by direct penetration in the region of elongation or through wounds that expose the vascular system. For infection, the roots have to be in close proximity to the pathogen, since hyphae that grow out of the germinating resting structures can traverse only a limited distance (≤2 mm).

Fig. 61. Bines affected by the lethal form of Verticillium wilt. Bine of a diseased plant early (left) and late (right) in the season. (Courtesy S. Radišek)

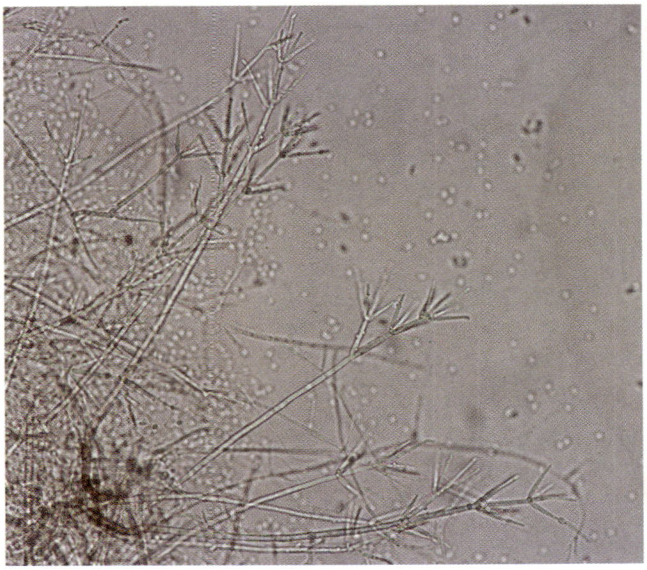

Fig. 62. Verticillate conidiophores of *Verticillium albo-atrum*. (Courtesy S. Radišek)

After colonization of the root cortex, infection leads to the invasion of the xylem tissue in which conidia are formed. The transpirational stream carries the conidia upward, where they are trapped in border pits or at vessel end walls. For colonization to continue, germ tubes, which penetrate adjacent vessel elements and begin a new infection cycle, must be produced. The accumulation of fungal biomass in the xylem disturbs the conduction of water and nutrients. Both fungi also produce phytotoxins and a variety of plant cell wall-degrading enzymes, including pectinases, polysaccharidases, and proteinases. As plant tissues die or senesce, resting structures begin to form as a result of necrotrophic development. These resting structures are released into the soil as the plant materials decompose and thus provide initial inoculum for the next disease cycle. *V. albo-atrum* has the ability to survive in soil free of host plants for 2–4 years, and *V. dahliae* can survive without a host for 15 years or longer.

Dissemination in hop yards occurs primarily through soil cultivation and movement of infected hop bines and leaves. A number of weeds common in hop yards act as symptomless carriers and provide a further means of spreading the pathogen and aid in the survival of the fungi in the soil. Inoculum densities tend to increase from year to year when susceptible cultivars are planted. Long-distance dispersal of the pathogens may occur in infected planting material and by the movement of hop waste from contaminated hop yards to disease-free areas.

Management

Disease management of Verticillium wilt requires an integrated strategy. There is no effective chemical control available, so crop rotation, planting resistant or tolerant cultivars, and carrying out phytosanitary measures are the only successful means of suppressing this disease. The eradication of affected plants and neighboring plants is necessary to restrict disease spread and slow the buildup of soil inoculum in infested hop yards. This measure is especially important in the case of lethal outbreaks in order to prolong the cropping life of an infested hop yard. Modified cultural practices that incorporate weed control, reduced cultivation, and limiting nitrogen application have proven to be important means of reducing disease incidence. Even when a contaminated hop yard is inspected regularly, many symptomless infected plants may escape detection and be harvested for machine picking. Hop waste thus provides the potential for further spread and should be subjected to thermal decomposition (burned or composted). In order to compost mechanically crushed hop bines and leaves, a pile approximately 2 m in height and width and should be formed. Such a dumping pile should be covered with plastic immediately to ensure warming on the surface of the pile and to prevent crop debris from being scattered by wind. After 2 months, such compost can be taken away to a field in which no host plants are produced.

Crop rotation is one of the cultural techniques recommended for reducing inoculum in the soil. Hop is a perennial plant that needs an extensive and permanent support system for cultivation, so performing crop rotation as a soil sanitation measure is impractical in hop production. A minimum of 4 years of crop rotation with nonhost plants (e.g., wheat or corn) or grass pasture can eliminate or reduce the inoculum of *V. albo-atrum* in hop yards affected by lethal wilt. Only resistant cultivars are suitable for replanting such hop yards. Although resistant cultivars present the most effective way to control this disease, no hop genotypes have been discovered that are immune to Verticillium wilt. Even highly resistant cultivars can become infected and often show mild or no symptoms. Both fungi can be carried undetected by planting material. Only hop yards that are free of Verticillium wilt should therefore be included in the production of planting material.

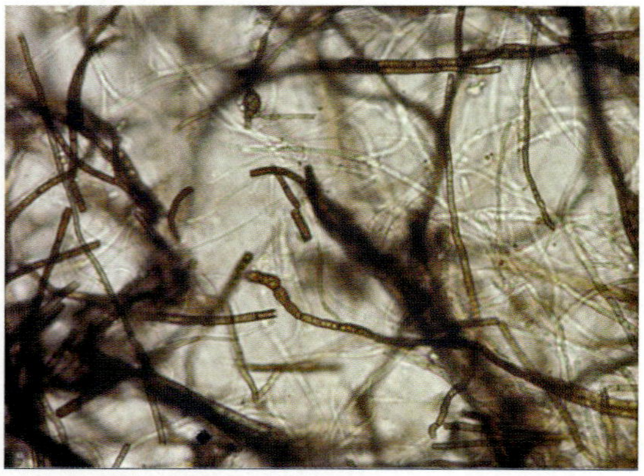

Fig. 63. Dark resting mycelium of *Verticillium albo-atrum.* (Courtesy S. Radišek)

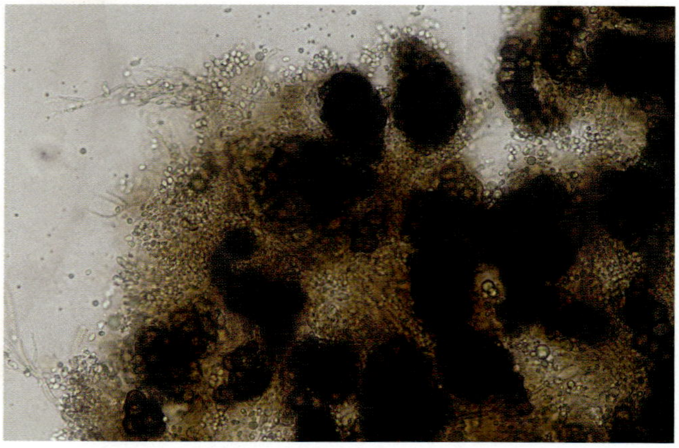

Fig. 64. Microsclerotia of *Verticillium dahliae.* (Courtesy S. Radišek)

Selected References

Clarkson, J. M., and Heale, J. B. 1985. Pathogenicity and colonization studies on wild-type and auxotrophic isolates of *Verticillium albo-atrum* from hop. Plant Pathol. 34:119-128.

Engelhard, A. W. 1957. Host index of *Verticillium albo-atrum* Reinke & Berth. (including *Verticillium dahliae* Kleb.). Plant Dis. Rep. Suppl. 244:23-49.

Radišek, S., Jakše, J., Simončič, A., and Javornik, B. 2003. Genetic variability and virulence among *Verticillium albo-atrum* isolates from hop. Eur. J. Plant Pathol. 116:301-314.

Sewell, G. W. F. 1959. Direct observation of *Verticillium albo-atrum* in soil. Trans. Brit. Mycol. Soc. 42:312-321.

Sewell, G. W. F., and Wilson, J. F. 1966. *Verticillium* wilt of the hop: The survival of *Verticillium albo-atrum* in soil. Ann. Appl. Biol. 58:241-249.

Sewell, G. W. F., and Wilson, J. F. 1967. *Verticillium* wilt of the hop: Field studies on wilt in a resistant cultivar in relation to nitrogen fertilizer applications. Ann. Appl. Biol. 59:265-273.

Sewell, G. W. F., and Wilson, J. F. 1984. The nature and distribution of *Verticillium albo-atrum* strains highly pathogenic to the hop. Plant Pathol. 33:39-51.

Talboys, P. W. 1987. Verticillium wilt in English hops: Retrospect and prospect. Can. J. Plant Pathol. 9:68-77.

(Prepared by S. Radišek)

Fungal Diseases and Pathogens of Minor Importance

Several fungi are reported to attack hop infrequently, including *Cercospora cantuariensis*, *Ceriospora dubyi* Niessl,

Cladosporium spp., *Cylindrosporium humuli* Ellis & Everh., *Glomerella cingulata* (Stoneman) Spauld. & H. Schrenk, *Macrosporium* spp., *Phoma* spp., *Phyllosticta* spp., *Pythium* spp., *Rosellinia necatrix*, and *Synchytrium aureum* J. Schröt. Most of these organisms are ubiquitous in the environment and live saprophytically on decaying plant debris and diverse substrates. The diseases they cause on hop are uncommon and are often associated with senescent tissues or an attenuated host.

Alternaria Blight

Alternaria humuli E. G. Simmons was isolated from diseased leaves of hop in Alsace, France, in 1998. Disease symptoms were not reported, so it is not clear if this fungus attacks cones. Young cultures of *A. humuli* are subhyaline and arachnoid to loosely wooly and sporulate profusely on potato carrot agar. Distinct clusters of sporulation are apparent in 7-day-old colonies. Clusters contain simple to slightly branching conidiophores 40–50 × 3–4 µm. Primary conidia are often larger than secondary conidia and are elliptical to ovoid, are dull gray-tan with darker septa, and measure 28–38 × 10–11 µm. Secondary conidia are ellipsoid or broadly ovoid and 19–26 × 10–13 µm and have three to four transepta and one longiseptum in one to three of the transverse segments.

Black Mold

Black mold is caused by an unidentified *Cladosporium* sp. that causes a brown discoloration of bracts, giving affected cones a variegated or striped appearance. Dark spores of the fungus are apparent on affected bracts under low magnification. The discoloration is most prominent on cones protected from direct sunlight, such as those on low lateral branches. The disease causes negligible damage, but black mold is easily confused with downy mildew and misdiagnosis may lead to premature harvest in an attempt to minimize losses from downy mildew. *Cladosporium* spp. are also involved in the sooty mold complex.

Cercospora Leaf Spot

Cercospora cantuariensis Salmon & Wormald (syn. *Mycocentrospora cantuariensis* Deighton) is reported to cause circular to irregularly shaped leaf spots on hop in Austria, Germany, Slovenia, the United Kingdom, China, Korea, and Russia. Other *Cercospora* spp. (*C. humuli* Hori, syn. *C. humuli-japonici* Sawada, and *C. humuligena* Y. L. Guo & L. Xu) have been reported to infect leaves of various *Humulus* spp. in China, Japan, and Taiwan. Lesions on leaves are gray to white and have dark, purplish-brown margins surrounded by a diffuse yellow halo (Fig. 65). As with most leaf spot diseases, lesions may coalesce and become necrotic under conditions favorable to the disease. In such conditions, cones also can be affected, showing irregular brown necrotic areas on the bracts and bracteoles (Fig. 66). Conidia are pale brown and cylindrical with tapering ends. They have five to 19 septa and are approximately 14 × 200 µm (Fig. 67). The disease causes minimal damage.

Fig. 65. Gray to white lesions with dark margins caused by infection of leaves by *Cercospora cantuariensis*. (Courtesy S. Radišek)

Hop-Drop

Macrosporium spp. have been implicated in a disease known as hop-drop that has been reported infrequently in the United Kingdom. The cone-bearing stalks become brown and discolored and eventually detach from the plant as cones near maturity, causing a considerable number of cones to fall from the plant. The symptoms of the disease are conspicuous and may be alarming, but actual yield losses are minimal.

Phoma Wilt

Phoma exigua var. *exigua* Desm. has been reported to cause a canker and wilt disease on hop in continental Europe, New

Fig. 66. Reddish-brown discoloration of the tips of bracts and bracteoles caused by infection of cones by *Cercospora cantuariensis*. (Courtesy S. Radišek)

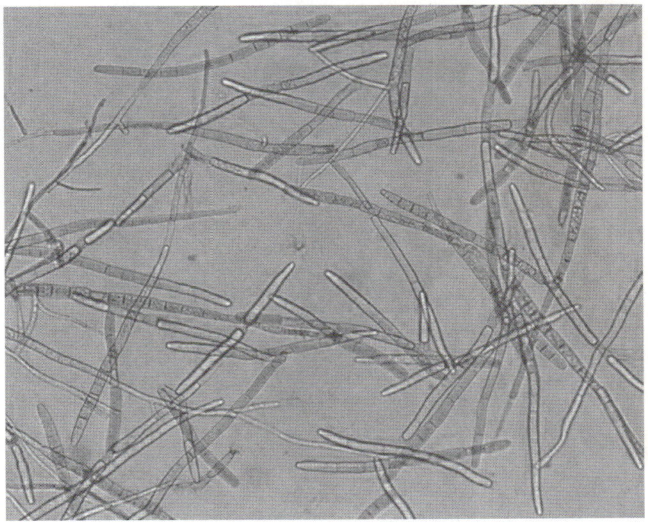

Fig. 67. Septate conidia of *Cercospora cantuariensis* (×200). (Courtesy S. Radišek)

Zealand, and China. *P. exigua* has also been reported to cause a leaf spot and cone necrosis disease in Slovenia. Disease symptoms of Phoma wilt first appear as chlorotic spots on bines but progress into gray-brown, elongated lesions (Fig. 68). It causes browning on cones (Fig. 69) that can result in poor cone qualtiy and crop rejection. Plants are stunted and unthrifty and may wilt completely. Pycnidia on affected bines are dark, small (80–260 µm), and spherical. Conidia are hyaline, nonseptate, nonguttulate, ellipsoid, and 4.5–5.5 × 1.5–2.5 µm (Fig. 70). The pathogen survives in and on infested crop debris and decaying organic matter and is dispersed by splashing water and wind. *P. exigua* var. *exigua* has also been implicated in diseases on more than 35 hosts.

Fig. 68. *Phoma exigua*-infected leaves showing oval, grayish white lesions. (Courtesy of S. Radišek).

Fig. 69. Dark reddish-brown lesions on hop cones caused by *Phoma exigua*. (Courtesy of S. Radišek).

Rhizoctonia solani

Rhizoctonia solani J. G. Kühn (teleomorph *Thanatephorus cucumeris* (A. B. Frank) Donk) has been reported, in rare instances, to cause lesions on young shoots of the Brewers Gold cultivar in British Columbia. Lesions are sunken and reddish brown to black. Affected shoots are stunted and may collapse if they become girdled by a lesion near the crown. *R. solani* has also been implicated in root and wilt disease complexes in continental Europe, although the pathogen is presumed to be a minor component of these complexes.

Rosellinia (Dematophora) Root Rot

Rosellinia necatrix Prill. (anamorph: *Dematophora necatrix* R. Hartig) attacks numerous forest and orchard trees, shrubs, and vines in temperate regions throughout the world. Affected plants are stunted and wilted, and they may senesce prematurely and die. A white mass of fungal hyphae forms on and in roots, and prominent brown to black perithecia (1–2 mm in diameter) appear on the diseased tissues. The pathogen spreads primarily through dead and decaying roots in the soil and is often disseminated by the movement of infested soil and infected ornamentals.

Selected References

Boerema, G. H., de Gruyter, J., Noordeloos, M. E., and Hamers, M. E. C. 2004. *Phoma* Identification Manual. Differentiation of Specific and Infra-Specific Taxa in Culture. CABI, Wallingford, UK.

Chupp, C. 1953. A monograph of the fungus genus *Cercospora*. (Published by the author)

Jasnić, S., Đurić, T., and Sabo, J. 1998. *Rhizoctonia solani*, parazit hmelja u Jugoslaviji. Zašt. bilja 225:245-255.

Jones, W. 1952. Infection of basal shoots of hop plants by *Rhizoctonia solani*. Sci. Agric. 32:114-115.

Radišek, S., Leskosek, G., Jakše, J., and Javornik, B. 2008. Occurrence of *Cercospora cantuariensis* on hop in Austria and Slovenia. New Dis. Rep. (accepted).

Radišek, S., Jakše, J., Javornik, B., and de Gruyter, J. 2007. First report of *Phoma exigua* as a pathogen of hop in Slovenia. Plant Pathol. 57:381.

Salmon, E. S., and Ware, W. M. 1936. The Cladosporium disease of the hop. J. S. E. Agric. Coll. Wye. 38:53-54.

Salmon, E. S., and Wormald, H. 1923. A new *Cercospora* on *Humulus*. J. Bot. (Lond.) 61:134-136.

Salmon, E. S., and Wormald, H. 1923. Three new diseases of the hop. J. Minist. Agric. 30:430-435.

Simmons, E. G. 2002. *Alternaria* themes and variations (305-309) *Lewia*/*Alternaria* revisited. Mycotaxon 83:127-145.

(Prepared by D. H. Gent and S. Radišek)

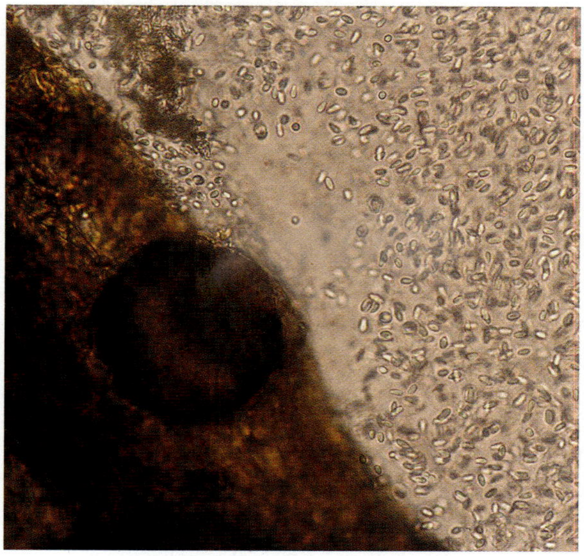

Fig. 70. Pycnidium and conidia of *Phoma exigua* (×200). (Courtesy of S. Radišek).

Diseases Caused by Viruses and Viroids

Apple Fruit Crinkle

Apple fruit crinkle viroid (AFCVd) was first reported in 1976 from apples showing fruit crinkle symptoms in an orchard in Iwate Prefecture, Japan. A new strain, AFCVd-hop, was reported in 2004 from hop in Akita Prefecture, Japan, where hop stunt-like symptoms were prevalent.

Symptoms

AFCVd-hop was epidemic in several hop yards in Taiyu town, Akita Prefecture. Yard inspections revealed a sporadic distribution of hop plants with stunting (Fig. 71) and severe leaf curling in the upper part of the bines (Fig. 72). Slight stunting was first observed during the fourth and fifth growing seasons, reducing yield by 20–25% compared with noninfected plants. Alpha-acid content was also reduced to half that of the noninfected plants in these years. The ratio of alpha- to beta-acid contents of cones harvested from severely infected plants was 1:1 or less compared with 2:1 in noninfected plants. The symptoms resemble those incited by *Hop stunt viroid.*

Causal Agent

AFCVd-hop consists of an infectious, circular, single-stranded RNA molecule with 368–372 nucleotides. AFCVd-hop shares 93–98% sequence similarity with AFCVd isolates from apple and 85–87% with *Australian grapevine viroid* (AGVd) isolates from grapevine. Routine diagnosis of AFCVd by nucleic acid hybridization or reverse transcription-polymerase chain reaction (RT-PCR) assays is essential for effective detection and management.

Disease Cycle and Epidemiology

Little is known about the disease cycle and epidemiology of AFCVd in hop. The viroid is sap-transmissible. Once the pathogen is introduced into a yard, it is readily spread mechanically by hands and cutting tools and by contact with bines through farmers' practices such as crowning, stripping, and picking.

Fig. 71. Stunted hop plants in a yard severely affected by a hop strain of *Apple fruit crinkle viroid.* (Courtesy T. Sano)

Fig. 72. Severe leaf curling in the upper part of bines infected with a hop strain of *Apple fruit crinkle viroid.* (Courtesy T. Sano)

Management

It is essential to inspect plants in yards at regular intervals. Infected plants should be destroyed completely, including the roots. While the epidemic was active in Japan, 10 plants on either side of any infected plant in the same row and three plants in the adjacent rows were destroyed, even though the plants were not symptomatic. Once an infected plant was detected in a yard, growers were encouraged to change or disinfect farm equipment and cutting tools between hop yards and plants. Growers are encouraged to replace diseased plants, or preferably all the plants in the yard, with virus- and viroid-free plants.

Selected References

Pethybridge, S. J., Hay, F. S., Barbara, D. J., Eastwell, K. C., and Wilson, C. R. 2008. Viruses and viroids infecting hop: Significance, epidemiology, and management. Plant Dis. 92:324-338.

Sano, T., Yoshida, H , Goshono, M., Monma, T., Kawasaki, H., and Ishizaki, K. 2004. Characterization of a new viroid strain from hops: Evidence for viroid speciation by isolation in different host species. J. Gen. Plant Pathol. 70:181-187.

(Prepared by T. Sano)

Apple Mosaic

Apple mosaic virus (ApMV), formerly *Prunus necrotic ringspot virus* (PNRSV), occurs worldwide and is an important and common pathogen of hop. The first report of infection of hop by an ilarvirus was from the successful transmission of a virus from hop to cucumber (*Cucumis sativus* L.). Prior to the establishment of virus certification schemes, ApMV was prevalent throughout commercial yards in the United States, the United Kingdom, Japan, the Czech Republic, Germany, Slovenia, Belgium, France, Romania, Yugoslavia, China, Australia, and New Zealand. The incidence of ApMV in yards established with material certified free from ApMV varies with yard age and cultivar susceptibility.

Symptoms

ApMV may induce chlorotic ringspots that can become necrotic (Figs. 73–75) or form oak-leaf line patterns on leaves (Figs. 76 and 77). Symptom expression appears to be influenced by growing conditions and resultant virus titer; symptomless infections are also common. ApMV is generally regarded as having the greatest impact on yield among the commonly occurring viruses of hop. Infection by ApMV may reduce cone weight by up to 50% and alpha-acid content by up to 10% in sensitive cultivars, but yield and alpha-acid losses vary markedly depending on the cultivar, location, and season. Certain cultivars appear tolerant of ApMV, although co-infections with other common viruses such as *Hop mosaic virus* may act synergistically to reduce yield. In addition to production losses, ApMV can decrease the survival of propagated hop softwood cuttings and emergence following first dormancy, increasing the cost of yard establishment.

Causal Agent

ApMV, a member of subgroup III of the *Ilarvirus* genus in the family *Bromoviridae*, consists of three pleomorphic isometric particles 25–29 nm in diameter. Each particle contains one component of the tripartite single-stranded RNA genome of approximately 2.1, 3.0, and 3.5 kb. RNA 1 contains two open reading frames (ORFs) coding for the coat protein (25 kDa) and the cell-to-cell movement protein (32.3 kDa). RNA 2 has

Fig. 73. Chlorotic rings on leaves infected by *Apple mosaic virus* in the cultivar Columbus. (Courtesy K. C. Eastwell)

Fig. 74. Chlorotic rings on leaves infected by *Apple mosaic virus* in cultivar Opal. (Reprinted, by permission, from Pethybridge et al., 2008)

Fig. 75. Chlorosis and ringspot symptoms caused by *Apple mosaic virus* in cultivar Nugget. (Courtesy D. H. Gent)

Fig. 76. Ringspots caused by *Apple mosaic virus* can become necrotic and merge to produce an oak-leaf pattern on affected hop plants. Symptoms pictured here are in cultivar Willamette. (Reprinted, by permission, from Pethybridge et al., 2008)

a single ORF coding for the viral polymerase (99.9 kDa), and RNA 3 also has a single ORF coding for a putative viral methyltransferase/helicase (118.4 kDa). All three RNA molecules and the coat protein are required for infection.

ApMV infections in hop are routinely detected by serological testing, primarily enzyme-linked immunosorbent assay (ELISA) or reverse transcription-polymerase chain reaction (RT-PCR) assays. Fluctuations in ApMV titer in infected plants, particularly during hot and dry growing conditions, can result in occasional failure to detect infections with ELISA. *Cucumis sativus* is a suitable assay and propagation host, exhibiting chlorotic local lesions, systemic chlorosis, and stunting.

Serological typing of ilarvirus isolates from hop has demonstrated two distinct serotypes, one resembling ApMV and referred to as ApMV-Hop (ApMV-H) (synonymous with "PNRSV-A", "A", and "HP-1"), and a second that is intermediate between ApMV and PNRSV called ApMV-Intermediate (ApMV-I) (syn. "PNRSV-I", "I", "C", and "HP-2"). Biophysical differences between intermediate and PNRSV isolates from hosts of the genus *Prunus*, the failure of intermediate isolates to react with cRNA probes prepared against PNRSV-cherry isolates, and RNA sequence analysis of the coat protein coding region of a range of isolates of both serotypes from hop have clearly shown that both the apple and intermediate serotypes are strains of ApMV and not PNRSV. There is a report of a "cherry" serotype (most likely PNRSV) infecting wild or feral hop plants in Germany.

Disease Cycle and Epidemiology

ApMV has a wide host range. It infects plants in at least nine families. ApMV has no known vector, and spread within hop yards is primarily to adjacent plants, most likely through mechanical transmission or root grafting. Spread in hop yards is generally slow, varying according to cultivar susceptibility. Efficient transmission of ApMV between hop plants following shoot and root contact, including simulated pruning and root grafts, has been demonstrated experimentally. Plant-to-plant contact through intertwining of young plants and basal growth as well as cultural operations such as mowing, stringing, training, leaf stripping, and thinning may be involved in transmission. Some evidence also exists for the transmission of ApMV by root grafting within commercial hop yards. This phenomenon occurs naturally between many woody perennial plant species.

Fig. 77. Oak-leaf pattern and ringspots caused by *Apple mosaic virus* in cultivar Nugget. (Courtesy D. H. Gent)

The role of pollen is not considered important in the movement of ApMV between hop plants. This conclusion is made on the basis of the observation that there has been no reduction in virus spread following conversion to seedless hops and the resulting decline in the numbers of male plants and thus contaminated pollen. Also, virus transmission occurs in both flowering and deflowered plants at similar frequencies. In some cultivars in yards where males are present, up to 51% of seeds have been infected by ApMV. However, the role of seed transmission in the spread of ApMV in commercial yards is of little consequence because planting material is produced through vegetative propagation. Moreover, many modern cultivars are triploid and hence sterile.

Management

Establishment of yards with material certified to be virus-free and the strategic use of herbicides to eliminate volunteer plants are the current means of control in hop. Yard reestablishment is costly, and, where possible, the benefits need to be weighed against the expense, since individual cultivars can vary markedly in both their susceptibility and tolerance to infection. However, determining the economic benefit of yard replacements (both in yield gains from virus freedom and the expected duration of such gains) requires prior data on the relative reinfection rate following establishment with virus-free material and yield (cone and brewing acid) reductions following ApMV infection for each cultivar, which may not be readily available.

The effectiveness of alternative practices such as herbicide application for basal shoot control has not been quantified but may reduce virus spread because of reduced sap transfer. This may be a practical management option to reduce virus spread.

Selected References

Crosslin, J. M., and Mink, G. I. 1992. Biophysical differences among Prunus necrotic ringspot ilarviruses. Phytopathology 82:200-206.

Crowle, D. R., Pethybridge, S. J., Leggett, G. W., Sherriff, L. J., and Wilson, C. R. 2003. Diversity of the coat protein-coding region among Ilarvirus isolates infecting hop in Australia. Plant Pathol. 52: 655-662.

Eppler, A., and Dahdahbiglou, R. 1991. Experiments on the non-vectorial transmission of PNRV in hops. Med. Fac. Landbouww. Rijksuniv. Gent. 56:577-587.

Pethybridge, S. J., Nelson, M. E., Eastwell, K. C., Klein, R. E., Kenny, S. T., and Wilson, C. R. 2002. Incidence and spatial distribution of viruses in hop gardens of Washington State. Plant Dis. 86:661-665.

Pethybridge, S. J., Wilson, C. R., Sherriff, L. J., Leggett, G. W., and Munro, D. 2000. Virus incidence in Australian hop (*Humulus lupulus* L.) gardens and varietal field resistance to virus infection. Aust. J. Agric. Res. 51:685-690.

Pethybridge, S. J., Wilson, C. R., Hay, F. S., Leggett, G. W., and Sherriff, L. J. 2002. Mechanical transmission of apple mosaic ilarvirus in Australian hop (*Humulus lupulus*) gardens. Ann. Appl. Biol. 141:77-85.

Shiel, P. J., Alrefai, R. H., Domier, L. L., Korban, S. S., and Berger, P. H. 2000. The complete nucleotide sequence of apple mosaic virus RNA3. Arch. Virol. 140:1247-1256.

Smith, D. R., and Skotland, C. B. 1986. Host range and serology of Prunus necrotic ringspot virus serotypes isolated from hops (*Humulus lupulus*) in Washington. Plant Dis. 70:1019-1023.

(Prepared by C. R. Wilson, S. J. Pethybridge, and K. C. Eastwell)

Arabis Mosaic

The hop strain of *Arabis mosaic virus* (ArMV-H) has been reported on hop in a number of countries, including Belgium, the Czech Republic, the United Kingdom, France, Germany, South

Africa, and New Zealand, but it has not been found in Japan, Romania, or the United States. In Australia and the United States, there were early reports of diseases now known to be caused by ArMV-H in cultivars imported from the United Kingdom, but in recent years there have been no reports of ArMV-H in hop in these countries. It is thought that a widespread shift to locally bred material and the restricted distribution of the nematode vector, *Xiphinema diversicaudatum* (Micoletzky) Thorne, may have led to the eventual eradication of this virus from hop yards in these countries. However, in New Zealand and South Africa, ArMV-H was transmitted from early imported cultivars to locally bred cultivars and became endemic. The nematode vector is indigenous to the United Kingdom, and the differing experiences in Australia and other countries where ArMV-H was introduced presumably depended on whether the vector also was introduced. Infection by ArMV-H can significantly reduce yield in susceptible cultivars. In the cultivar Bullion in the United Kingdom, diseased plants yielded 23–26% less than healthy plants and alpha-acid content in healthy plants was 7.1% (by weight) compared with 6.5% in diseased plants. Yield losses of 50% from infection by ArMV-H in the cultivar Fuggle have been reported.

Fig. 78. Curved, weak shoots with bare-bine symptom caused by *Arabis mosaic virus.* (Courtesy D. J. Barbara)

Symptoms

Infection by ArMV-H has been associated with several diseases, including bare-bine or spidery hop, split leaf blotch, nettlehead, and hop chlorotic disease. Symptoms of bare-bine or spidery hop are evident early in the season. Plants produce fewer and weaker shoots that develop a curvature and form darker and smaller leaves than healthy plants (Fig. 78). Plants often recover after the weak shoots are removed during training, but they can develop split leaf blotch (Fig. 79) or nettlehead (Fig. 80) later in the season. Split leaf blotch is manifested as translucent, yellow, oily blotches between the leaf veins that often split as the leaf expands. Early work showed this condition was graft-transmissible, and it was long assumed to be of viral etiology. Less severe symptoms were noted on other cultivars and were termed "slight blotch." However, the association between slight blotch and ArMV-H is yet to be proven.

Early in the season, plants exhibit symptoms of bare-bine before developing stiff, erect shoots with shortened internodes that fail to climb and fall away from strings. Leaves display an upward rolling of the margins and often develop vein clearing or mottle, with enations on the underside. Symptom expression may be variable. Symptoms can be present on only a few shoots, and affected shoots or entire plants may express symptoms one year but not the next. Symptoms are inconspicuous in shoots produced in hot weather. Similarly, cuttings taken from affected plants can give rise to apparently healthy plants in a glasshouse. In Australia, nettlehead can cause a reduction in both the number and weight of cones, which is partially related to a reduced seed set as a result of delayed flowering until after the time of maximum pollen availability. Nettlehead is consistently associated with ArMV-H infection, but not all plants infected by ArMV-H develop nettlehead. Nettlehead is associated with the presence of ArMV-H and a satellite nucleic acid (SNA), and symptom severity is related to the amount and particular SNA present.

Hop chlorotic disease was described in the United Kingdom during the 1930s and was also reported in Germany, Russia, and Poland, but it occurs only sporadically. The disease causes severe distortion and chlorosis of leaves produced early in the season (Fig. 81). Leaves often assume a characteristic "parrot-beak" shape. The disease is confined to lower parts of the bine.

Fig. 79. Translucent, yellow, oily blotches between leaf veins characteristic of split leaf blotch symptoms caused by *Arabis mosaic virus.* (Courtesy D. J. Barbara)

Fig. 80. Weak bines falling away from strings, a symptom associated with nettlehead, caused by *Arabis mosaic virus.* (Courtesy D. J. Barbara)

Causal Agent

ArMV is a nepovirus, named after the host plant *Arabis hirsuta* L. in which it was originally detected. Particles are polyhedral, approximately 28–30 nm in diameter, and have a single capsid protein with an electrophoretically estimated molecular weight of about 55,000 Da (505 amino acids from viral genome sequence). ArMV sediments into three components: a top component (50S) of protein shells devoid of RNA, a middle component (90–120S) containing 27–40% single-stranded RNA (RNA-2) by weight with a molecular weight of 1.4 to 2.2 $\times 10^6$ Da (depending on the isolate), and a bottom component (120–130S) containing 42–46% single-stranded RNA (RNA-1) with a molecular weight of approximately 2.4×10^6 Da. The nepovirus genome is bipartite, and both RNA-1 and RNA-2 are necessary for normal infection and multiplication. Like other isolates of the type strain, ArMV-H is associated with satellite nucleic acids (SNAs). The SNA associated with nettlehead disease is a 300-nucleotide, covalently closed, single-stranded RNA with an apparent molecular weight by electrophoresis of 75,000 Da. The SNA is encapsidated in the coat protein of ArMV-H, and it occurs in both middle and bottom components during sedimentation. When present, it usually makes up about 80% of the nucleic acids of purified preparations. This SNA is not necessary for infection or replication of ArMV-H and has no sequence homology with the helper virus genome, but it depends upon it for its own multiplication. Multiplication of SNA is supported only by the ArMV-H strain of ArMV. A larger (1,092-nucleotide) SNA has been reported, but its significance in relation to disease is not known.

In comparison to ArMV-H, other strains of ArMV have wider host ranges. ArMV from strawberry is unable to infect hop and vice versa. The type strains of ArMV and ArMV-H can be differentiated by mechanical inoculation to *Chenopodium amaranticolor* (Coste & A. Reyn.) Coste & Reyn. and *C. quinoa* Willd. ArMV causes chlorotic local lesions and systemic chlorotic mottling, and ArMV-H causes mild systemic vein netting. ArMV-H can also be distinguished from other strains serologically.

Disease Cycle and Epidemiology

The dagger nematode *X. diversicaudatum* (Fig. 82) is considered the main vector of ArMV, including hop strains. Other nematode vectors reported include *X. coxi* Tarjan, *X. bakeri* Williams, *X. index* Thorne & Allen, *Longidorus caespiticola* Hooper, and *Paralongidorus maximus* Bütschli, but evidence for transmission of ArMV by these nematodes is considered insufficient. *X. diversicaudatum* is also a vector of *Strawberry latent ringspot virus*, which has been reported in hop in the former Czechoslovakia. *X. diversicaudatum* is more efficient at transmitting hop strains of ArMV than those from strawberry. Although hop generally supports low population densities of *X. diversicaudatum*, the root system of hop can occupy several cubic meters, so even a low density of nematodes can represent many thousands per plant, with significant potential for virus transmission. *X. diversicaudatum* is widespread throughout Europe and occurs sporadically in parts of Australia, New Zealand, North America, and South Africa. In the United Kingdom, *X. diversicaudatum* occurs most frequently in arable, permanent pasture and deciduous woodland sites. It is widespread in the southern United Kingdom, where hop production areas are now concentrated.

X. diversicaudatum is a relatively long-lived nematode with a life cycle of up to 3 years from hatching through four juvenile stages and differentiation into adults. There are reports of individual females living for more than 5 years. Information on longevity on hop is lacking. *X. diversicaudatum* can live, but not multiply, in soil without higher plants for 3 years. Each of the four juvenile stages and adults can be found throughout the year, but there can be an increase in the proportion of mature females during seasonal flushes of new plant growth. The rate of movement of *X. diversicaudatum* under field conditions has been estimated at 0.3–0.7 m per year.

Transmission of ArMV by *X. diversicaudatum* occurs after a minimum access period of 24 h. Adults and juveniles of *X. diversicaudatum* transmit ArMV with equal efficiency, and nematodes can remain viruliferous for approximately 15 months in soil without higher plants. The virus is not retained through the molt between stages of the life cycle and is not passed from the female through the egg.

The distribution of virus-infected plants arising from nematode transmission is often aggregated, corresponding to the distribution of the vector within the soil. However, the distribution of infected plants is also affected by the health of the crop at planting and by the distribution of viruliferous nematodes at planting. Sites where the vector is absent remain free of the virus, unless introduced with planting material, in which case there is a tendency for a random distribution with limited secondary spread. At sites with viruliferous vectors, spread occurs quickly to plants within areas infested by the vector, leading to an aggregated distribution of infected plants. The incidence of ArMV-H in a hop yard in the United Kingdom increased from 2.9 to 43.6% in 7 years and to 61.8% in 12 years following the planting ArMV-H-infected materials in a yard where *X. diversicaudatum* was widely distributed.

Early reports of nettlehead disease in hop in the United Kingdom indicated that the worst outbreaks occurred in hop rows situated alongside existing or former hedgerows and after plowing of permanent pasture. Because ArMV-H has a restricted host range, these outbreaks are unlikely to be caused by nematode transmission between infected plants within hedgerows and hop plants. Instead, they were likely caused by the presence of suitable hosts and high populations of *X. diversicaudatum* in such areas. *X. diversicaudatum* occurs rarely in hop yards in southern Germany and then only in one region, despite the nematode's common presence in pastures surrounding hop yards and in hedges and woodlands. This absence has been attributed to the high level of cultivation and soil disturbance for weed control within yards.

Fig. 81. Distortion and chlorosis of leaves with hop chlorotic disease, caused by *Arabis mosaic virus.* (Reprinted, by permission, from Pethybridge et al., 2008)

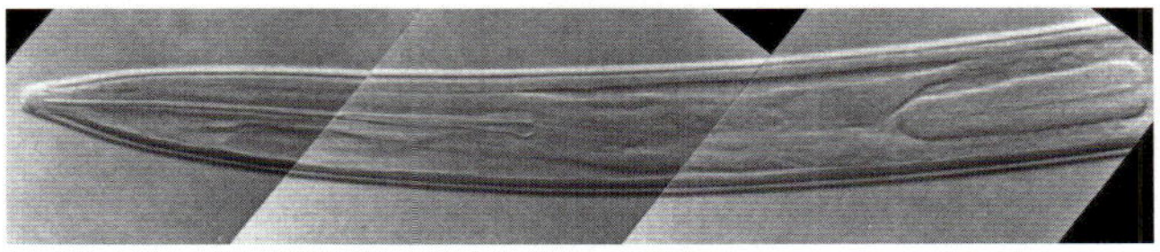

Fig. 82. Anterior portion of *Xiphinema diversicaudatum*, nematode vector of *Arabis mosaic virus.* (Courtesy F. S. Hay)

The pathogen can be transmitted by graft inoculation and sap inoculation and through seed, but mechanisms of transmission other than nematode vectors are considered of little importance in hop yards. However, with strains of ArMV other than ArMV-H, dissemination of infected crop and weed seed may be important.

Management

The main method of managing ArMV-H is the use of certified planting stock and a fallow period prior to replanting. In the United Kingdom, if a site has no history of virus or disease, it is recommended that virus-free material be planted immediately, irrespective of the presence of vector nematodes. Alternatively, if a site has a history of virus or disease, then a 2-year fallow or rotation to a nonhost (e.g., grass or strawberry) is recommended prior to planting, irrespective of the presence of nematode vectors. To ensure successful control, old plants should be removed to eliminate any subsequent regrowth.

The use of nematicides alone is ineffective in controlling the reinfection of healthy planting stock with ArMV-H. The inability of nematicides to eradicate ArMV-H has been attributed to insufficient movement through heavy soils to depths that hop roots and vector nematodes can penetrate and to the efficiency of virus transmission by even low vector population densities. However, application of nematicides in conjunction with a 20-month or longer fallow can reduce ArMV-H to undetectable levels.

Selected References

Adams, A. N., Barbara, D. J., and Davies, D. L. 1987. The etiology of hop chlorotic disease. Ann. Appl. Biol. 111:365-371.

Brown, D. F. J., and Coiro, M. I. 1983. The total reproductive capacity and longevity of individual female *Xiphinema diversicaudatum* (Nematoda: Dorylaimida). Nematol. Mediterr. 11:87-92.

Davies, D. L., and Clark, M. F. 1983. A satellite-like nucleic acid of arabis mosaic virus associated with hop nettlehead disease. Ann. Appl. Biol. 103:439-448.

Hay, F. S., and Close, R. C. 1992. Distribution of *Xiphinema diversicaudatum* (Micoletzky, 1927) Thorne, 1939 in commercial hop (*Humulus lupulus* L.) gardens in New Zealand and implications for the spread of arabis mosaic virus. New Zeal. J. Crop Hort. 20:367-370.

Kaper, J. M., Tousignant, M. E., and Steger, G. 1988. Ribonucleotide sequence predicts circularity and self-cleavage of 300-ribonucleotide satellite of arabis mosaic virus. Biochem. Bioph. Res. Co. 154:318-325.

Pethybridge, S. J., Hay, F. S., Barbara, D. J., Eastwell, K. C., and Wilson, C. R. 2008. Viruses and viroids infecting hop: Significance, epidemiology, and management. Plant Dis. 92:324-338.

Pitcher, R. S., and McNamara, D. G. 1973. The control of *Xiphinema diversicaudatum*, the vector of arabis mosaic virus in hops. Ann. Appl. Biol. 75:468-469.

Thresh, J. M., and Pitcher, R. S. 1978. The spread of nettlehead and related virus diseases of hop. Pages 291-298 in: Plant Disease Epidemiology. P. R. Scott and A. Bainbridge, eds. Blackwell Scientific, Oxford.

Valdez, R. B., McNamara, D. G., Ormerod, P. J., Pitcher, R. S., and Thresh, J. M. 1974. Transmission of the hop strain of arabis mosaic virus by *Xiphinema diversicaudatum*. Ann. Appl. Biol. 76:113-122.

(Prepared by F. S. Hay and D. J. Barbara)

American hop latent virus, *Hop latent virus*, and *Hop mosaic virus*

Three viruses of the genus *Carlavirus* are known to infect hop plants. These are *Hop mosaic virus* (HpMV), *Hop latent virus* (HpLV), and *American hop latent virus* (AHLV). Hop mosaic disease, caused by HpMV, was first recognized in 1923. The viral nature of the disease was finally confirmed and described in 1967. HpLV was first described in 1966, and AHLV was first detected in 1976. Mixed infections of these carlaviruses are common, but the viruses can be distinguished readily by serological means and by their experimental host range. HpMV and HpLV occur worldwide, while AHLV has been found in the United States and New Zealand. In Australia, AHLV was recorded in post-entry quarantine and subsequently eliminated.

Symptoms

On most current commercially important cultivars of hop plants, infection by any one or more of these carlaviruses is symptomless. Lack of symptom expression is a frequent characteristic of infection by carlaviruses.

The most striking symptom of HpMV on sensitive cultivars is the development of chlorotic mosaic; the chlorotic lesions often develop bounded by major leaf veins. In addition, the bines climb poorly, often failing to attach to supports. Adding to the appearance of unthrifty growth is the development of shortened internodes. The hop cultivars that typically display conspicuous symptoms, such as chlorotic flecking, in response to infection by HpMV are those of the Golding type or cultivars derived from them (Figs. 83 and 84). Sensitive cultivars

Fig. 83. Interveinal chlorosis of leaves caused by *Hop mosaic virus* on a sensitive Golding-type hop plant. Symptoms can range from mild (right) to severe (left). (Reprinted, by permission, from Pethybridge et al., 2008)

Fig. 84. Chlorotic mosaic symptoms caused by *Hop mosaic virus* in cultivar J78. (Courtesy P. McGee)

yield poorly and die prematurely when infected. HpMV caused significant crop losses in areas where HpMV-sensitive cultivars were grown; however, sensitive cultivars presently constitute a small percentage of commercial hop production. HpMV does not induce symptoms on any of the limited plants in its experimental host range other than sensitive hop cultivars. A useful propagation host for HpMV is *Nicotiana occidentalis* H.-M. Wheeler.

HpLV does not cause symptoms on any commercial hop cultivars. However, chlorotic flecks have been induced on seedlings derived from the cultivar Cluster. Although HpLV is able to infect several species of the genus *Chenopodium*, there is no reliable or diagnostic symptom development on these plants.

AHLV does not induce acute symptoms on any of the hop cultivars that have been tested. AHLV can be differentiated from the other carlaviruses that infect hop plants because it has a broader host range and induces reliable symptom expression on the experimental hosts *Chenopodium quinoa* Willd. and *Datura stramonium* L. On these herbaceous hosts, symptoms include chlorotic or necrotic lesions and systemic vein banding and rings. HpLV and HpMV also infect *C. quinoa*, but they do not reliably induce recognizable symptoms.

The impact of carlavirus infections on hop growth, cone yield, and levels of brewing organic acids varies markedly with the cultivar. Many hop cultivars appear tolerant of carlaviruses, but sensitive cultivars have shown decreases in cone yield by up to 62%, 70%, and 14% and decreases in alpha-acid content by up to 18%, 44%, and 12% for HpMV, HpLV, and AHLV, respectively. In general, the consequences of infection by these carlaviruses on hop cone brewing characteristics appear to mimic changes associated with advanced maturity.

Causal Agent

Carlaviruses are characterized by slightly flexuous, filamentous, rod-shaped virions (Fig. 85). The genomes of carlaviruses consist of a single molecule of single-stranded, positive sense RNA. The three carlaviruses that infect hop are typical in these respects.

The virions of HpMV with particle dimensions of 651 × 13.8 nm are noticeably shorter than those of the other carlaviruses that infect hop plants. The size of the genome, estimated through gel electrophoresis, is 8,800 nucleotides. The virus particle is encapsidated by multiple copies of a single protein, each with a calculated molecular mass of 34.1 kDa. When hop mosaic disease was first recognized, the isolates of HpMV that were identified were lethal to Golding types of hop plants. Since that time, additional isolates of HpMV have been described, some of which induce a similar pattern of leaf chlorosis but with no dramatic reduction in growth or increased mortality. The mild strains of HpMV may be widely distributed in hop production regions of the world. No additional information is available regarding the molecular or serological relationships of the different strains of HpMV.

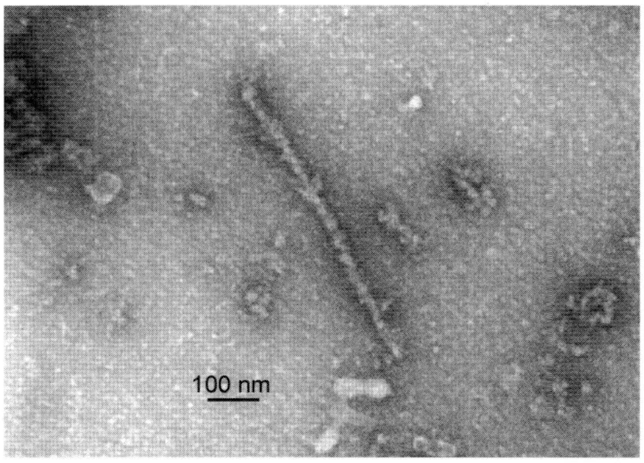

Fig. 85. Filamentous carlavirus virions visualized with an electron microscope. (Courtesy D. H. Gent)

Isolated virions of HpLV are 674 × 14.4 nm. The genome of HpLV consists of 8,612 nucleotides in one single-stranded RNA species. The predicted molecular mass of the capsid protein is 34.3 kDa

The virions of AHLV are also filamentous and 678 × 15 nm in size. The single-stranded RNA genome consists of 8,605 nucleotides plus a polyadenylated tract at the 3′ terminus. The genome encodes a capsid protein that has a molecular mass of 34.5 kDa, determined using the predicted amino acid sequence.

Although AHLV is a member of the *Carlavirus* genus, it is serologically distinct from the other carlaviruses that infect hop plants. Whereas HpLV and HpMV seem to share distant antigenic similarities with each other and many other carlaviruses, AHLV is only remotely related serologically to other carlaviruses, if at all. Immunosorbent electron microscopy revealed a mild reaction of AHLV with antisera prepared to *Nerine latent carlavirus* (NeLV).

Disease Cycle and Epidemiology

HpMV, HpLV, and AHLV are transmitted in a nonpersistent manner by the damson-hop aphid, *Phorodon humuli* (Schrank), and it is presumed that this is the major route of natural transmission in the field in areas where this aphid is present. In addition, HpMV and HpLV transmission by the potato peach aphid, *Macrosiphum euphorbiae* (Thomas), and the green peach aphid, *Myzus persicae* (Sulzer), commonly found within hop yards has been demonstrated. Experimental aphid transmission of these carlaviruses is inefficient, although the natural spread of HpMV can be quite rapid in hop yards. In spite of the shared insect vectors, several surveys indicate that AHLV infection generally occurs less often than that of the other carlaviruses.

A small percentage of annual stinging nettle, *Urtica urens* L., has been found to be naturally infected by HpMV within hop yards in the United Kingdom. The role of this alternate host in virus epidemiology is not known.

There is no evidence of seed or pollen transmission. Vegetative propagation of virus-infected plants is a major route of dissemination of HpMV, HpLV, and AHLV. This aspect of virus epidemiology is controlled by certification programs that generate virus-tested plant material for further propagation.

Carlaviruses are also transmitted readily by mechanical inoculation under experimental conditions, which likely plays a role in plant-to-plant spread in the field. Similarly, root grafting has been implicated as a mechanism of local spread.

Management

Consistent with most carlaviruses, those that infect hop plants have a fairly limited host range. Very few host plants naturally occur in or around hop yards. Since they are transmitted by the damson-hop aphid, it is difficult to achieve complete control of the viruses in areas where virus and vectors occur. The carlaviruses are latent, causing relatively slight reductions in yield in most commercial hop cultivars. Therefore, they are generally tolerated in hop yards. The major exception is HpMV in sensitive cultivars. Nevertheless, planting or replanting missing plants with virus-tested plants is a good strategy. The impact of carlaviruses may be greatest during the establishment phase of hop yards, because the negative impact on production is less pronounced in subsequent years. This was demonstrated for cultivars Nugget, Opal, Pride of Ringwood, and Victoria. Moreover, it was shown that HpMV and HpLV significantly increased mortality of cuttings during propagation, with HpMV having the greater effect.

Selected References

Adams, A. N., and Barbara, D. J. 1983. Host range, purification and some properties of two carlaviruses from hop (*Humulus lupulus*): Hop latent and American hop latent. Ann. Appl. Biol. 101:483-494.

Eppler, A. 1989. Sensitivity, tolerance and resistance in the hop–CARLA-virus relationship. Pages 99-108 in: Proc. Int. Workshop

Hop Virus Dis. Giessen. Eugen Ulmer GmbH & Co., Stuttgart, Germany.
Hay, F. S., Close, R. C., and Beatson, R. 1989. The effect of virus infections on the yield and alpha/beta acid content of the hop variety Super alpha. Pages 137-148 in: Proc. Int. Workshop Hop Virus Dis. Giessen. Eugen Ulmer GmbH & Co., Stuttgart, Germany.
Kanno, Y., Iida, H., Yoshikawa, N., and Takahashi, T. 1994. Some properties of Hop mosaic virus isolated in Japan. Ann. Phytopathol. Soc. Jpn. 60:675-680.
Pethybridge, S. J., Hay, F. S., Barbara, D. J., Eastwell, K. C., and Wilson, C. R. 2008. Viruses and viroids infecting hop: Significance, epidemiology, and management. Plant Dis. 92:324-338.
Pethybridge, S. J., Wilson, C. R., Hay, F. S., Leggett, G. W., and Sherriff, L. J. 2002. Effect of viruses on agronomic and brewing characteristics of four hop (*Humulus lupulus*) cultivars in Australia. Ann. Appl. Biol. 140:97-105.
Probasco, G., and Murphey, J. M. 1996. The effects of hop viruses on brewing and agronomic characteristics in the hop variety Chinook. Master Brewers Assoc. Am. Tech. Q. 33:160-165.

(Prepared by K. C. Eastwell, S. J. Pethybridge, and C. R. Wilson)

Humulus japonicus latent virus (*Humulus japonicus virus*)

Humulus japonicus latent virus (HJLV) was originally called *Humulus japonicus virus,* but it was renamed in response to amendments in the International Committee on the Taxonomy of Viruses rules for naming plant viruses. HJLV was isolated and described from *Humulus japonicus* Siebold & Zucc. grown from seed imported into the United Kingdom from China. Subsequently, HJLV was detected commonly by serological testing in *H. lupulus* L. and *H. japonicus* near Beijing. All infected *H. japonicus* plants in the United Kingdom were destroyed, and hence the virus is believed to have been eliminated from that country. Little testing has been conducted in other parts of the world, and the geographical range of this virus remains unclear. However, the introduction of HJLV into the United Kingdom through propagation material highlights the need for quarantine procedures to prevent the spread of viruses and other diseases among hop-growing regions of the world.

Symptoms

No symptoms have been associated with infection by HJLV in *H. japonicus* or *H. lupulus.* Following mechanical inoculation in the United Kingdom, infection of *H. lupulus* was localized to inoculated leaves. However, testing of both wild *H. japonicus* and commercial *H. lupulus* in China suggested infection was systemic.

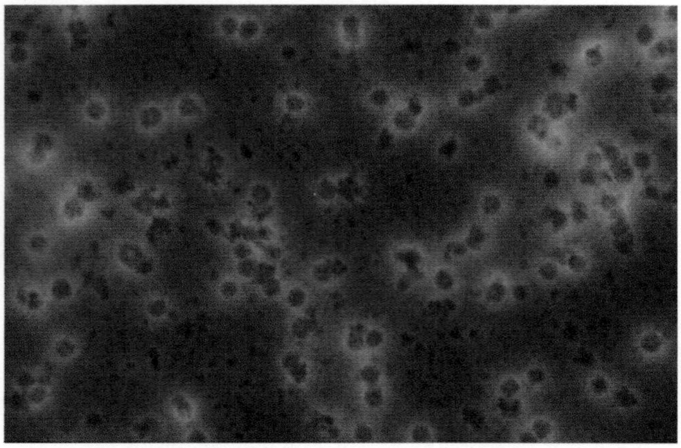

Fig. 86. Nucleocapsids of *Humulus japonicus latent virus.* (Courtesy D. J. Barbara)

Causal Agent

HJLV is a member of subgroup 3 of the *Ilarvirus* genus in the family *Bromoviridae.* Viruses in the subgroup are related serologically, but data from the full genome sequence suggests HJLV is divergent from the other subgroup members. Nucleocapsids are quasi-isometric and vary in size (Fig. 86). They generally range from 24 to 33 nm in diameter and have an obvious arrangement of the surface capsomer. Virions contain four segments of single-stranded RNA of 3,382, 2,810, 1,938, and 987 bp (initially estimated electrophoretically at 1.31, 1.05, 0.75, and 0.39 kDa). One virion structural protein has been identified and was initially estimated by electrophoresis as being 26.3 kDa, but it was determined to be 27.1 kDa from sequence data. HJLV is distantly related serologically to *Prunus necrotic ringspot virus* but unrelated to *Prune dwarf virus, American plum line pattern virus, Tobacco streak virus,* and *Elm mottle virus.*

This virus is known to infect 27 plant species in 10 families, many of which are diagnostic for ilarviruses. *Chenopodium quinoa* Willd. and *C. amaranticolor* (Coste & A. Reyn.) Coste & Reyn. are suitable hosts for local lesion assays. Symptoms on inoculated *C. quinoa* leaves are pale green lesions that become irregular, sunken, white or brown necrotic areas. Systemic symptoms vary but include leaf distortion, chlorotic vein clearing, mottling, flecks and rings, and chlorotic or necrotic spots. Symptoms on inoculated *C. amaranticolor* leaves are pinpoint, white, necrotic lesions, while systemic infections are characterized by chlorotic spots and rings. Symptoms of infection on *Cucumis sativus* L. (cultivar Bedfordshire Prize Ridge) are large, slightly pale green lesions on inoculated leaves, systemic vein clearing, and mosaic or mottle, leaf distortion, and stunting. In inoculated *Nicotiana clevelandii* A. Gray leaves, large, usually incomplete, etched necrotic ringspots occur locally, while systemic symptoms are chlorotic mottling and necrotic spots and ringspots. *C. quinoa, C. amaranticolor,* and *N. clevelandii* are suitable diagnostic hosts. *C. quinoa* is a suitable maintenance and propagation host. Diagnosis is best confirmed by serological or molecular methods since symptom expression on test plants can be confused with that of *Apple mosaic virus,* which also occurs commonly in hop. Immunosorbent electron microscopy and decoration have also been used effectively for diagnostic purposes.

Disease Cycle and Epidemiology

HJLV can be transmitted by mechanical inoculation to a variety of different hosts including *H. japonicus* and *H. lupulus.* Seed transmission has been demonstrated in *H. japonicus* and infrequently in *C. quinoa.* Other ilarviruses are transmitted through their association with pollen and nonspecific mechanical inoculation via feeding activity of thrips. HJLV has been associated with the pollen of *C. quinoa.* Most virus particles are borne externally as with other ilarviruses such as *Prunus necrotic ringspot virus* and *Tobacco streak virus.* The symptomless infection of *H. japonicus* and *H. lupulus* is of concern since production losses from this virus are unknown.

Management

The restricted geographical range and low incidence of HJLV means that management techniques have not been necessary outside China. It is unknown what control measures, if any, have been taken in China, where HJLV occurs frequently in both wild *H. japonicus* and cultivated *H. lupulus,* at least in the Beijing area. For HJLV, it is likely that adherence to strict quarantine protocols applicable to each country will prevent introduction. If necessary, management would be similar to protocols followed for many other viruses, including the identification and roguing of infected plants and planting of certified virus-free material.

Selected References

Adams, A. N., Clark, M. F., and Barbara, D. J. 1989. Host range, purification and some properties of a new ilarvirus from *Humulus japonicus*. Ann. Appl. Biol. 114:497-508.

Fauquet, C. M., Mayo, M. A., Maniloff, J., Desselberger, U., and Ball, L. A. 2005. Virus Taxonomy. Eighth Report of the International Committee on Taxonomy of Viruses. Elsevier Academic Press, San Diego, CA.

Pethybridge, S. J., Hay, F. S., Barbara, D. J., Eastwell, K. C., and Wilson, C. R. 2008. Viruses and viroids infecting hop: Significance, epidemiology, and management. Plant Dis. 92:324-338.

Scott, S. W., and Zimmerman, M. T. 2006. The complete sequence of the genome of *Humulus japonicus latent virus*. Arch. Virol. 151:1683-1687.

(Prepared by S. J. Pethybridge and D. J. Barbara)

Hop latent viroid

Following the description of *Hop stunt viroid* (HpSVd) in 1977, a second viroid infecting hop was detected and described as *Hop latent viroid* (HpLVd) in 1988. The "hop viroid-like RNA fast, HV-f" reported in 1987 was most likely also HpLVd. HpLVd was probably not detected earlier because of its apparent latency, the inability to mechanically transmit the viroid to common herbaceous indicator species, and sequence differences between HpLVd and HpSVd. HpLVd has been reported from hop-production areas in the United Kingdom, Germany, New Zealand, Korea, Poland, Slovenia, Japan, the Czech Republic, the United States, and Australia and Brazil in material recently imported from the United States and Australia. In the majority of these areas, the incidence of HpLVd is considered high or ubiquitous, although an early survey may have overestimated the occurrence in some countries. Infection by HpLVd is deleterious to production in some cultivars.

Symptoms

Infection by HpLVd is symptomless in most cultivars. However, in the United Kingdom, infections in cultivar Omega are associated with chlorosis, slow growth, and the production of fewer laterals with smaller cones (Fig. 87). The effects of HpLVd infection are less marked in two other cultivars in the United Kingdom, Wye Northdown and Wye Challenger, but it may still reduce alpha-acid production by approximately 20% and increase myrcene levels (Fig. 88). Infection by HpLVd also reduces rooting of softwood cuttings and establishment.

Fig. 87. Diminished growth and leaf chlorosis caused by *Hop latent viroid* infection of cultivar Omega; healthy plant in center and *Hop latent viroid*-infected plants behind. (Courtesy D. J. Barbara)

Causal Agent

HpLVd consists of a single-stranded, closed, circular RNA molecule 256 nucleotides long that has been fully sequenced (Genbank accession X07397) and that, in common with most viroids in the family *Pospiviroidae*, is thought to normally adopt a stable, rod-like, secondary structure. The genome contains an upper and a lower conserved central region and a series of conserved sequence motifs. HpLVd is classified as a distinct species within the genus *Cocadviroid* because of low sequence similarity to other viroids and other sequence features. The sequence of HpLVd is quite dissimilar to that of *Hop stunt viroid* (HpSVd), which is in the genus *Hostuviroid* of the same family. Although viroids generally form complex populations of natural sequence variants (quasispecies), no sequence variants have been reported for HpLVd in natural isolates. However, quasispecies have been induced in heat-treated isolates and likely exist in natural populations.

HpLVd has a very limited natural host range of only three species: *Humulus lupulus* L., *H. japonicus* Siebold & Zucc., and *Urtica dioica* L. (stinging nettle). This pathogen can be detected by dot blot hybridization and reverse transcription-polymerase chain reaction (RT-PCR) assays. Probes have been developed from a HpLVd clone for which both isotopic and nonisotopic labeling have been used for detection in dot blots. The dot blot technique is recommended for ease of use and rapid processing of large numbers of samples. However, RT-PCR assays are approximately 10,000 times more sensitive than dot blot hybridization.

Spatial and temporal fluctuations in HpLVd titers within the hop plant, which vary according to cultivar and region, can complicate detection. HpLVd can be detected only the year after mechanical inoculation in early spring. Early in the season, titers vary in the aerial parts of the plant and are generally higher near the base, but toward the end of the growing season, HpLVd is usually present throughout infected plants. HpLVd titers are also higher in mericlones grown under in

Fig. 88. Infection by *Hop latent viroid* results in no apparent reduction in vigor or other visible symptoms in cultivar Wye Northdown. *Hop latent viroid*-infected plant marked by the black bag at the base. (Reprinted, by permission, from Pethybridge et al, 2008)

vitro conditions and in roots than they are in the leaves of field-grown plants. HpLVd titer can vary temporally, but findings have been inconsistent. In the Czech Republic and Germany, titers decreased in summer compared with early spring, but in the United Kingdom, the opposite trend was found. This difference may reflect the generally higher temperatures in these parts of continental Europe in summer compared with the United Kingdom. HpLVd detection is often unreliable in early spring because of low titers, but in practice, reliable detection can be achieved by selecting any tissue type from low on the bines in late spring onward. Despite the fact that HpLVd titers are high in roots, these tissue are not used for routine detection because of sampling difficulties.

Disease Cycle and Epidemiology

Plants can become infected quickly after yard establishment. Spread of HpLVd within yards is predominantly by mechanical means, such as harvesting and mechanical cultural operations. Thus, the majority of new infections occur in plants that are adjacent within rows to those that are infected by HpLVd. However, in the United Kingdom, new infections, arising some distance from infected plants, suggest the role of an unknown vector. Physical cutting with infected instruments is more efficient for transmission than plant-to-plant contact. Any cultural operation that initiates wounds likely contributes to spread, but because production practices vary between countries and cultivars, the relative contribution of different cultural operations also varies. Pollen transmission has not been found. Reports of seed transmission are mixed, varying between none in the United Kingdom to 8% transmission efficiency in the Czech Republic. The main means of introduction into commercial yards is likely through infected vegetatively propagated planting material. No evidence has been found for spread by insects through transencapsidation in co-infecting viruses. *Urtica dioica* (stinging nettle) has been reported as a natural host of HpLVd, but its role in the epidemiology of the disease is presumed to be minor.

Management

The effect of HpLVd on yield has been great enough to include it in hop certification schemes. Selection of healthy planting material is pivotal to management. Given the high incidence of this pathogen in some cultivars or countries, selection of viroid-free material may be impossible and therapy may be required to rid the material of HpLVd. Both thermotherapy and cold treatment can reduce viroid titer, but in either case, excision and culturing of meristematic tissue is required after treatment. In the United Kingdom, HpLVd-free plants were produced successfully following storage at 2–4°C in the dark for periods of 8 and 21 months, followed by excision of explants less than 0.5 mm long. Thermotherapy can also significantly reduce HpLVd titer and lead to viroid elimination. Optimal reductions are obtained using cycles of heat treatment consisting of exposure to 36°C for 14 days. However, the original titer can be restored following cultivation under normal conditions for 6 months or longer. Reduction in viroid titer following heat treatment is correlated with the induction of a nucleolytic complex, which cleaves viroid RNA and hence leads to viroid degradation. Whether heat or cold therapy is used to reduce viroid titer before mericloning, verification of freedom from HpLVd by RT-PCR following winter dormancy is recommended.

Selected References

Adams, A. N., Barbara, D. J., and Morton, A. 1991. Effects of hop latent viroid on weight and quality of the cones of the hop cultivar Wye Challenger. Ann. Appl. Biol. 118:126-127.

Adams, A. N., Morton, A., Barbara, D. J., and Ridout, M. S. 1992. The distribution and spread of hop latent viroid within two commercial plantings of hop (*Humulus lupulus*). Ann. Appl. Biol. 121:585-592.

Barbara, D. J., Morton, A., Adams, A. N., and Green, C. P. 1990. Some effects of hop latent viroid on two cultivars of hop (*Humulus lupulus*) in the UK. Ann. Appl. Biol. 117:359-366.

Cajza, M., Wypijewski, K., and Pospiesny, H. 1996. The use of reverse transcription and polymerase chain reaction (RT-PCR) for the detection of hop latent viroid (HpLVd). J. Plant Prot. Res. 37:52-58.

Matoušek, J., and Patzak, J. 2000. A low transmissibility of hop latent viroid (HpLVd) through a generative phase of hop (*Humulus lupulus* L.). Biol. Plant. 43:145-148.

Morton, A., Barbara, D. J., and Adams, A. N. 1993. The distribution of hop latent viroid within plants of *Humulus lupulus* and attempts to obtain viroid-free plants. Ann. Appl. Biol. 123:47-53.

Pethybridge, S. J., Hay, F. S., Barbara, D. J., Eastwell, K. C., and Wilson, C. R. 2008. Viruses and viroids infecting hop: Significance, epidemiology, and management. Plant Dis. 92:324-338.

Puchta, H., Ramm, K., and Sänger, H. L. 1988. The molecular structure of hop latent viroid (HLV), a new viroid occurring world-wide in hops. Nucleic Acids Res. 16:4197-4216.

Von Knabel, S., Seigner, E., and Wallnofer, P. R. 1999. Detection of hop latent viroid (HpLVd) using the polymerase chain reaction (PCR). Gesunde Pflanz. 51:234-239.

(Prepared by S. J. Pethybridge, D. J. Barbara, C. R. Wilson, and K. C. Eastwell)

Hop Stunt

Hop stunt is a serious disease of cultivated hop that was first recognized in the 1940s in Nagano and later in Fukushima prefecture of Japan. By the 1960s, the disease had spread throughout the major hop-producing areas of northern Japan. The causal agent was identified as a new viroid in 1977 and named *Hop stunt viroid* (HpSVd). The disease was found in South Korea in 1988 in rhizomes imported from Japan, but it is no longer found in the substantially reduced hop-production area in South Korea. The viroid was also confirmed in North America in 2004. The disease has not been widely described in other hop-growing regions of the world.

Symptoms

The degree of symptom development is cultivar and climate dependent. Since the viroid was limited primarily to hop production in Japan and South Korea, most published information relates to symptom expression in cultivars important in those regions, namely Kirin II and Shinshuwase. Typical symptoms observed in these cultivars include stunting (Figs. 89 and 90), leaf curling, and small cones. The height and average cone weight of severely infected plants are approximately 33% less than those of healthy plants. The total number of cones from infected bines may or may not be affected, but on a per dry weight basis, infected cones have about one third as much alpha-acid as cones from healthy plants. Cones of infected plants have approximately the same number of bracts, but in the cultivar Shinshuwase, the cones are narrower as a result of shortened bracts. The number of lupulin glands is reduced significantly in cones, although the number of glands on leaves is not affected. Plants affected by hop stunt may bloom 8–10 days before healthy plants.

Visible stunting of plants, for which the disease is named, becomes apparent only several years after the plants become infected. This symptom results from a shortening of the internodes of the main bine and of the lateral branches. The degree of stunting is temperature-dependent, and more severe stunting occurrs in warmer growing regions. Inhibition of the development of hooks or barbs on the bine hampers the ability of infected plants to climb, enhancing the appearance of stunting.

Foliar symptoms can be diagnostic. In Japan, infected leaves droop severely from the basal part of the petioles and the edges of the leaves curl downward. A pair of leaf petioles drooping downward and a pair of lateral shoots growing upward forms a characteristic X shape (Fig. 91). In commercial production in North America, the drooping leaves are less evident. However, yellow-green leaves are prominent among the basal foliage that develops early in the growing season (Fig. 92), and the development of lateral branches is inhibited (Fig. 93). On some sensitive cultivars, yellow speckling along the major leaf veins is evident (Fig. 94). The interaction of cultivar and environment on symptom expression is not well known at this time.

When established plants become infected, alpha-acid production is reduced before stunting appears after a latent period of 3–5 years. However, in plants derived from diseased plants, severe symptoms are evident in the first growing season.

Fig. 91. Drooped petioles infected with *Hop stunt viroid*; drooped petioles and lateral shoot growing upward forms X shape. (Courtesy T. Sano)

Fig. 89. Stunted hop plants with light green leaves in a yard severely infected with *Hop stunt viroid*. (Courtesy T. Sano)

Fig. 92. Early in the growing season, *Hop stunt viroid*-infected plants (right) exhibit reduced vigor and numerous chlorotic leaves compared with uninfected plants (left). (Courtesy K. C. Eastwell)

Fig. 90. Restricted development of lateral branches on hop bines infected with *Hop stunt viroid*. (Courtesy T. Sano)

Fig. 93. *Hop stunt viroid*-infected plants (left) exhibit reduced lateral branching compared to uninfected plants (right). (Courtesy K. C. Eastwell)

Causal Agent

Hop stunt is caused by the *Hop stunt viroid* (HpSVd), a member of the *Pospiviroidae* family. The genome of HpSVd consists of a covalently closed circular RNA molecule that contains 294–303 nucleotides.

A large number of HpSVd variants have now been isolated from several woody hosts, including almond, apricot, citrus, cucumber, grapevine, peach, plum, and pomegranate. Some natural hosts exhibit disease symptoms, e.g., plum and peach dapple (red blotches on fruit) and citrus cachexia (gumming, pegging, pitting, and general health decline), but the others do not develop symptoms. Five distinct sequence groups of HpSVd variants have emerged on the basis of the similarity of sequences. Hop isolates from northern Japan and two Korean isolates are closely related to some grapevine isolates. Grapevine subgroup 1 isolates of HpSVd induce typical hop stunt symptoms when inoculated on hop. It has been suggested that HpSVd may have originated in grapevine and spread to hop in Japan in the early twentieth century.

Disease Cycle and Epidemiology

HpSVd is usually introduced into a new planting through infected propagation material. The potential for this route of introduction is greatly enhanced because of the long latent period before conspicuous symptoms develop.

HpSVd has the widest host range of known viroids. However, HpSVd has not been found to naturally infect any of the weeds commonly associated with hop farms. There is no evidence of transmission by insect or nematode vectors or through seeds or pollen. The primary means of long-distance dissemination of HpSVd is through infected propagation material.

Observations from several locations indicate that the viroid tends to move along rows rather than across them. This suggests that infection from bines rubbing on the wire is an inefficient mode of transmission and also that farm operations may be a major contributing factor to the spread of this disease. Once HpSVd is established in a yard, the viroid is readily transmitted mechanically by workers, cutting tools, and equipment during cultural activities. The number of infected plants gradually increases over time, predominantly along rows. Local transmission within the yard occurs primarily with early-season operations such as pruning (to set stringing dates), thinning, and mechanical leaf stripping. HpSVd remains infectious for extended periods in plant sap, and procedures that simultaneously release sap and wound plants could spread HpSVd.

Fig. 94. Yellow speckling along the major leaf veins of the sensitive cultivar Glacier infected with *Hop stunt viroid.* (Courtesy D. H. Gent)

HpSVd can remain infectious in plant debris in the field for 3 months, but it is uncertain what role this potential source of inoculum plays in the spread of the viroid. The viroid can survive in roots for more than 1 year and can be spread to remote locations by distribution of contaminated propagation materials.

Management

The best strategy to manage hop stunt is to use planting material known to be free from HpSVd. Once HpSVd is confirmed in a planting, aggressive actions are needed to control further spread of infection. Because of the long latent period, removal of symptomatic plants may allow infected plants to remain in the hop yard, which then serve as sources of inoculum. Therefore, several plants adjacent to infected plants should also be removed. It is imperative to kill or remove as much root as possible in preparation for replanting, because infected volunteer plants can reinitiate infections in the replanted area. Replanting the entire yard with viroid-free propagation material is recommended. However, if this is not feasible, some modification of horticultural practices may help limit further movement of HpSVd between and within yards. Field operations should be performed in the infected yards after they have been done in HpSVd-free yards. Decontamination of equipment between fields is also encouraged. A solution containing 2% formaldehyde and 2% sodium hydroxide is effective but corrosive and potentially hazardous for workers. Many other disinfectants including 5–20% bleach (sodium hypochlorite), 5% calcium hypochlorite, 1% sodium hydroxide, 1% formaldehyde, or 6% hydrogen peroxide have been proposed, but results have been inconsistent. Exposure to one of these disinfectant solutions for 10 minutes has reduced infection by some viroids. Thorough washing of farm equipment to remove plant residue and sap may help reduce the likelihood of transmission to new yards.

The increased use of contact herbicides to set training dates and remove weak shoots in the spring is preferable to the use of mechanical mowers that may transmit the viroid. Similarly, removing basal vegetation late in the season by chemical rather than mechanical means also reduces the risk of transmission.

Initial discovery of HpSVd in a hop yard is usually associated with the development of disease symptoms. Testing procedures are available to help detect the viroid and to assist in removing infected plants. A cucumber bioassay is an effective means of identifying infected plants. Cucumber plants (*Cucumis sativus* L. cultivar Suyo) grown at 30°C become severely stunted following mechanical inoculation. Certain cucurbits also develop leaf epinasty and curling, followed by necrotic mottling. More rapid diagnosis can be achieved by molecular hybridization or RT-PCR assays.

Selected References

Pethybridge, S. J., Hay, F. S., Barbara, D. J., Eastwell, K. C., and Wilson, C. R. 2008. Viruses and viroids infecting hop: Significance, epidemiology, and management. Plant Dis. 92:324-338.

Sano, T. 2003. Hop stunt viroid. Pages 207-212 in: Viroids. A. Hadidi, R. Flores, J. W. Randles, and J. S. Semancik, eds. CSIRO Publishing, Collingwood, Australia.

Sano, T., and Shikata, E. 1989. Hop stunt disease. Pages 159-164 in: Proc. Int. Workshop Hop Virus Dis. Giessen. Eugen Ulmer GmbH & Co., Stuttgart, Germany.

Sasaki, M., and Shikata, E. 1977. On some properties of hop stunt disease agent, a viroid. Proc. Jpn. Acad. Ser. B Phys. Biol. Sci. 53:109-112.

Sasaki, M., Fukamizu, K., Yamamoto, K., Ozawa, T., Kurokawa, M., and Kagami, Y. 1989. Epidemiology and control of hop stunt disease. Pages 165-178 in: Proc. Int. Workshop Hop Virus Dis. Giessen. Eugen Ulmer GmbH & Co., Stuttgart, Germany.

Yaguchi, S., and Takahashi, T. 1984. Response of cucumber cultivars and other cucurbitaceous species to infection by Hop stunt viroid. Phytopathol. Z. 109:21-31.
Yaguchi, S., and Takahashi, T. 1984. Survival of Hop stunt viroid in the hop garden. Phytopathol. Z. 109:32-44.
Yamamoto, H., Kagami, Y., Kurokawa, M., Nishimura, S., Ukawa, S., and Kubo, S. 1973. Studies on hop stunt disease in Japan. Rep. Res. Lab. Kirin Brewery Co. Ltd. 16:49-62.

(Prepared by K. C. Eastwell and T. Sano)

Viruses of Minor Importance

Natural infections with a number of additional viruses have been reported on hop or related *Humulus* species. Those reported herein have been found in multiple commercial hop yards in one or more regions or have been rigorously confirmed at moderate incidence in one region.

Alfalfa mosaic virus (AlMV)

Alfalfa mosaic virus (AlMV) has been reported on hop in the former Czechoslovakia, while an AlMV-like virus has been reported from hop in China. Both viruses were characterized by transmission to biological indicators and electron microscopy. Serological testing showed the virus from the former Czechoslovakia reacted with an antiserum raised to a European isolate of AlMV from pea, but the virus from China was serologically distinct from a Chinese AlMV isolate from soybean. Many isolates of AlMV have wide host ranges and are transmitted by several aphid species in a nonpersistent manner, by mechanical means, and through seeds in many plant species. Seed transmission in hop has not been tested. Mechanical transmission of the Chinese virus to hop resulted in systemic chlorotic spots in cultivar Styrian. No information on the effect of these viruses on yield or quality of hop products was reported.

Cucumber mosaic virus (CMV)

Cucumber mosaic virus (CMV) has been reported in Romania on hop that showed distorted leaves and chlorotic spotting and in Germany and the United Kingdom on plants of *Humulus scandens* auct. (syn. *H. japonicus* Siebold & Zucc.) that exhibited mottled foliage. Virus identity was determined through reaction in biological indicators and by electron microscopy, in vitro properties, aphid transmissibility, or cross-protection studies. CMV has a very wide host range and is transmitted in a nonpersistent manner by a variety of aphid species including the damson-hop aphid (*Phorodon humuli* [Schrank]). In contrast to the Romanian report, the German isolate failed to transmit to cultivated hop by grafting. Infections were infrequent in all reports.

Petunia asteroid mosaic virus (syn. tomato bushy stunt virus-petunia strain; PeAMV)

Petunia asteroid mosaic virus (syn. tomato bushy stunt virus-petunia strain; PeAMV) has been reported on hop from the former Czechoslovakia with infections confirmed by transmission to biological indicators, electron microscopy, and serology. Numerous symptoms were associated with infected plants including deformed, asymmetric leaves with necrotic crinkles, ringspot and line patterns, yellow spotting, perforated leaves, and leaf yellowing, although these plants were likely infected with additional viral pathogens. PeAMV has a moderately wide host range, naturally infecting plant species from three to nine families, has no known vector, and is transmitted through plant contact and horticultural operations that transfer contaminated sap.

Strawberry latent ringspot virus (SLRSV)

Strawberry latent ringspot virus (SLRSV) has been detected frequently infecting hop, according to a report from the former Czechoslovakia. Infections were confirmed by testing with biological indicators and serology. SLRV has a wide host range and is transmitted by nematode vectors. No information on the effect of SLRV on yield or quality of hop is available.

Tobacco necrosis virus (TNV)

Tobacco necrosis virus (TNV) has been reported infecting hop roots in the former Czechoslovakia, Romania, and France, with an incidence of up to 28%. Infections were confirmed by transmission to biological indicator species, electron microscopy, and serology. Detection was most successful when hop sap derived from sprouting buds of underground stems or young vegetative runners was used. TNV has a wide host range and is transmitted in nature by the plant-pathogenic soilborne fungus *Olpidium brassicae* (Woronin) P. A. Dang. during fungal infection. Distinct symptoms associated with TNV infection on hop have not been described, nor have effects on yield or quality of hop products been determined.

Selected References

Albrechtova, L., Chod, J., Kriz, J., and Fencl, J. 1979. Nachweis des Apple mosaic virus und des Tobacco necrosis virus in Hopfen. Phytopathol. Z. 94:45-55.
Bock, K. R. 1966. Arabis mosaic and Prunus nectroic ringspot viruses in hop (*Humulus lupulus* L.). Ann. Appl. Biol. 57:131-140.
Chod, J., Jokes, M., and Novak, M. 1979. The electron-microscopic proof of Tobacco necrosis virus in hop plants. Biol. Plant. 21:152-153.
Novak, J. B., and Lanzova, J. 1976. Identification of Alfalfa mosaic virus and Tomato bushy stunt virus in hop (*Humulus lupulus* L.) and grapevine (*Vitis vinifera* subsp. *sativa* (DC./Hegi) plants in Czechoslovakia. Biol. Plant. 18:152-154.
Pethybridge, S. J., Hay, F. S., Barbara, D. J., Eastwell, K. C., and Wilson, C. R. 2008. Viruses and viroids infecting hop: Significance, epidemiology, and management. Plant Dis. 92:324-338.
Polak, J., and Svoboda, P. 1988. Preliminary results of screening hop varieties and clones for the presence of some hop viruses in Czechoslovakia using ELISA. Pages. 27-31 in: Proc. Int. Workshop Hop Virus Dis. Giessen. Eugen Ulmer GmbH & Co., Stuttgart, Germany.
Schmidt, H. E. 1967. New results of hop virus research. Pages 47-53 in: Proc. Conf. Czechoslovak Plant Virol. 6th. Olomouc, Czechoslovakia. Academia Publishing House of the Czechoslowak Academy of Sciences, Prague.
Schmidt, H. E., and Karl, E. 1968. Untersuchungen uber eine Scheckung am Japanischen Hopfen [*Humulus scandens* (Lour.) Merr.]. Phytopathol. Z. 62:272-278.
Yu, J., and Liu, Y. 1987. The occurrence of three viruses in hop *Humulus lupulus* L. in China. Plant Pathol. 36:38-44.

(Prepared by C. R. Wilson, S. J. Pethybridge, and F. S. Hay)

Diseases Caused by Nematodes

Cyst Nematode

The main plant parasitic nematode associated with hop worldwide is the cyst nematode, *Heterodera humuli* Filipjev. In the late nineteenth century, this nematode was associated with the hop disease nettlehead, but subsequent studies determined no relationship, and nettlehead was later attributed to infection by *Arabis mosaic virus*. *H. humuli* occurs in hop yards in many countries, including Australia, Canada, Germany, New Zealand, the United Kingdom, Ukraine, Spain, and the United States. A cyst nematode with morphology similar to that of *H. humuli* and parasitic on *Urtica dioica* L., but not hop, was known as the "nettle race" of *H. humuli*, but it has subsequently been described as a separate species (*Heterodera riparia* (Kazachenko)).

Symptoms

Symptoms associated with feeding by *H. humuli* in hop include reduced bine length and chlorosis of leaves. In pot experiments, the growth of the cultivar Wye Northdown was inhibited when the population of *H. humuli* eggs exceeded 500 per gram of soil. In greenhouse studies in the United States, *H. humuli* reduced plant height and fresh and dry weights of shoots and caused mineral deficiencies in the hop cultivar Cascade. *H. humuli* also is known to interact synergistically with the soilborne fungus *Verticillium albo-atrum* Reinke & Berthier (the cause of Verticillium wilt) to adversely affect plant growth and increase the severity of wilt symptoms.

Causal Organism

The life cycle of *H. humuli* is similar to that of other members of Heteroderinae. Eggs within the pear-shaped dead body of the adult female (cyst) remain viable for several years (Fig. 95). The juvenile nematodes (J1) undergo one molt within the egg to form second-stage juveniles (J2), which hatch in the spring in response to exudates from hop roots (Fig. 96). Egg hatch coincides with a time of rising soil temperatures and rapid root growth as the hop plant emerges from dormancy. The J2 penetrate the root just behind the root cap and migrate parallel to the central cylinder (stele), where they become sedentary and initiate the formation of a feeding site (syncytium) formed from partial cell wall degradation, leading to large coalesced cells containing several nuclei. The syncytium acts as a transfer cell from which the nematode obtains water and nutrients. The J2 undergo three further molts, including two additional juvenile stages (J3 and J4), to become adult females or males (Fig. 97). Females are cream-colored and pear-shaped and remain sedentary, while males become vermiform again and migrate to mate with the sedentary females. Up to several hundred eggs are produced internally within the female. The female then dies, and the cuticle darkens to brown or black and hardens to form a protective cyst around the egg mass. Cysts often remain attached to the root surface. Mean lengths (excluding neck) and widths of *H. humuli* cysts from a German population were 524 μm (338–632 μm) and 367 μm (214–498 μm), respectively, while the mean length of J2 was 422 μm (368–468 μm).

Disease Cycle and Epidemiology

H. humuli can reach high population densities, with a maximum of 1,520–2,800 J2 per 100 ml of soil (from the top 20 cm) obtained in the spring in Tasmania, Australia. Similar population densities were observed in studies in the United Kingdom, with a peak of approximately 300–600 *H. humuli* J2 per 100 ml of soil early in the season. *H. humuli* has been retrieved at depths of up to 45 cm in hop yards in the United Kingdom, and over half the nematodes recovered were from 0–15 cm, coinciding with the depth of the majority of fibrous hop roots. However, high numbers were recovered from depths of up to 60 cm in Russian yards.

Fig. 95. Cysts of hop cyst nematode containing eggs (×10). (Courtesy F. S. Hay)

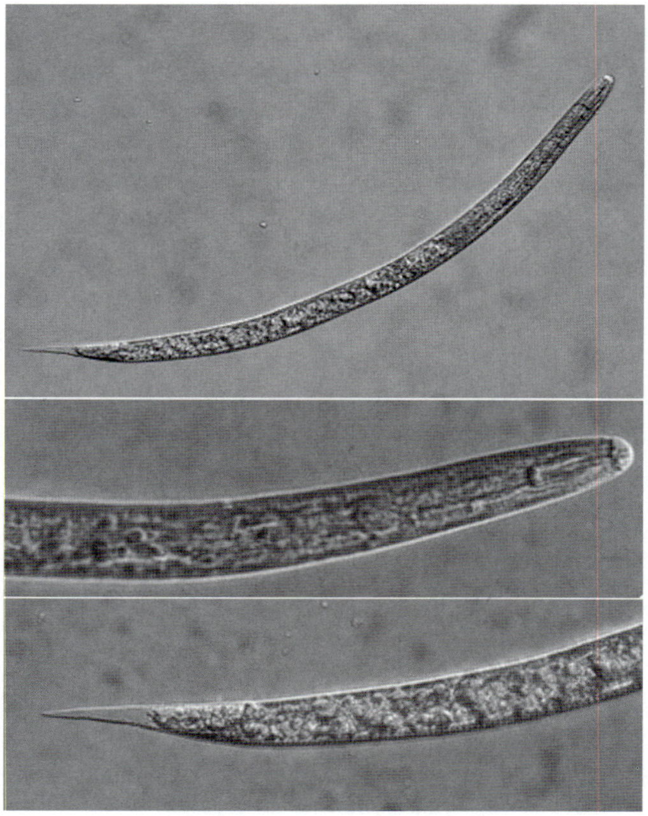

Fig. 96. Second-stage juvenile of *Heterodera humuli*. Top, entire second-stage juvenile (×100); middle, head region (×400); and bottom, tail region (×400). (Courtesy F. S. Hay)

Some research suggests that only one generation occurs annually. However, other work indicates there may be a second generation under appropriate conditions. In a study in the United Kingdom, numbers of J2 during the vegetative period of growth fluctuated in a pattern indicative of two or three generations, with a generation time of 50 days. However, in Australia, the numbers of J2 during the season is indicative of only one generation. Similarly, one generation was noted in the Kiev region of Russia in approximately 120 days.

Vermiform nematodes may also survive the winter. The presence of white females of *H. humuli* noted in hop yards in the United Kingdom in early spring was attributed to the development of overwintering J2 that had hatched the previous autumn, because soil temperatures during the winter would be expected to be too low to allow egg hatch. In Australia, the presence of males early in the season prior to shoot emergence from dormancy suggested that either some nematode development occurred during the winter or that males had survived from the previous season.

The perennial nature of hop and the size of its root system, together with its high growth rate during the spring, suggest that hop has a great capacity to tolerate nematode feeding. *H. humuli* has rarely been associated with reduced hop growth or yield under field conditions. However, *H. humuli* was associated with yield loss in the cultivar Pride of Ringwood in Australia. There was a 38% reduction in dry weight of hops per string between plants exposed to the highest population density of *H. humuli* J2 in mid-spring (10,080 per 100 ml) and the lowest population density (1,848 per 100 ml).

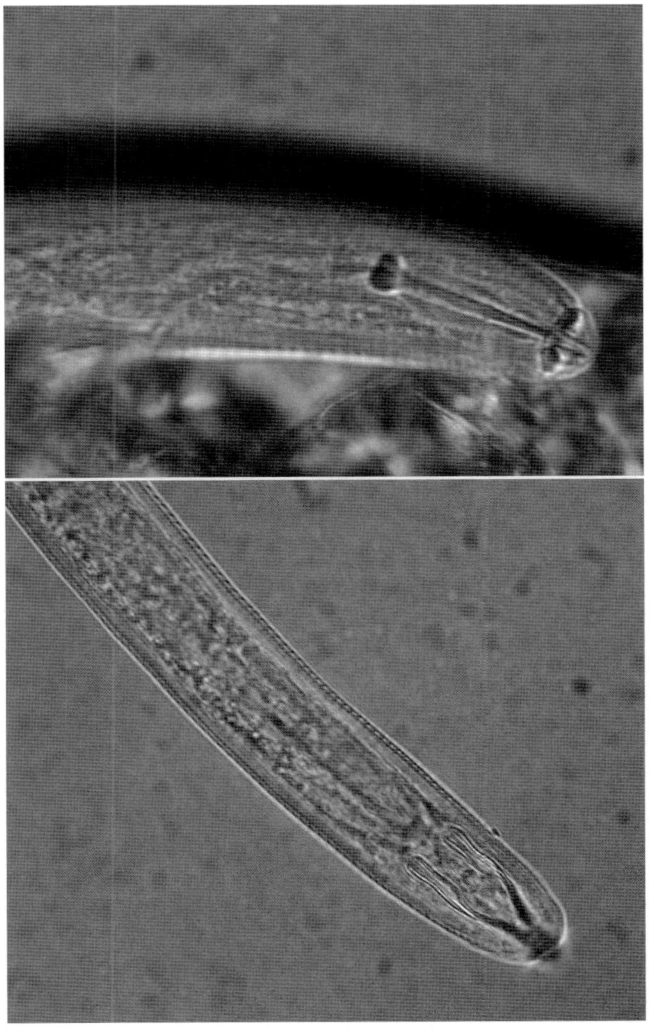

Fig. 97. Male *Heterodera humuli* (×400) showing head (top) and tail region (bottom) with spicules. (Courtesy F. S. Hay)

Management

Further studies are required to determine whether the effect of *H. humuli* on hop production is sufficient to warrant control measures. Control of nematodes by application of nematicides is unlikely to be an economic or effective control strategy because of the perennial nature of hop, the rapid multiplication rate of *H. humuli*, and the difficulty of applying effective doses of chemicals to the depths to which hop roots and nematodes can penetrate. However, there may be differences in the resistance or tolerance of hop cultivars that would provide an effective control strategy. Some research has also been undertaken on biological control of *H. humuli* with a nematophagous fungus (*Hirsutella rhossiliensis* Minter & B.L. Brady) identified in Bavaria, Germany, and several species of fungi associated with cysts in Ukraine found to be important in limiting nematode populations and controlling damage to hop.

Selected References

De Grisse, A., and Gillard, A. 1963. Morphology and biology of hop cyst eelworm (*Heterodera humuli* Filipjev 1934). Nematologica 9:41-48.

Hafez, S. L., Dorschner, K. W., and Hara, K. 1988. The effect of *Heterodera humuli* Filipjev on hops, *Humulus lupulus* cv. Cascade. J. Nematol. 20:89-94.

Hafez, S. L., Sundararaj, P., and Barbour, J. 1999. Impact of *Heterodera humuli* on growth and mineral nutrition composition of hops, *Humulus lupulus* cv. Cascade. Int. J. Nematol. 9:23-26.

Hay, F. S., and Pethybridge, S. J. 2003. Nematodes associated with hop production in Australia. J. Phytopathol. 151:369-375.

von Mende, N., and McNamara, D. G. 1995. Biology of the hop cyst nematode *Heterodera humuli*. I. Lifecycle and survival. Ann. Appl. Biol. 126:505-516.

von Mende, N., and McNamara, D. G. 1995. Biology of the hop cyst nematode *Heterodera humuli*. II. Host-parasite relationship of the nematode and its interactions with *Verticillium albo-atrum*. Ann. Appl. Biol. 126:517-526.

(Prepared by F. S. Hay and S. J. Pethybridge)

Other Nematode Species Associated with Hop

Many nematode species are found in association with hop plants, with 35 species recorded in hop yards in Ukraine, over 30 species in Bulgaria, more than 20 in Romania, and 14 from Russia. Where plant-parasitic nematodes have been identified, there often has been no attempt to determine which species are hosted by hop and which are hosted by weed species in hop yards. There has also been little effort to determine which species may be pathogenic to hop. Damage to hop by these nematode species is considered of minor importance; where damage has been noted, such cases have been of limited distribution. Certain nematodes are vectors of viruses or may predispose plants to infection by other pathogens. Their role as vectors or in facilitating infection by other pathogens is discussed in sections for other diseases (e.g., *Xiphinema* spp. associated with *Arabis mosaic virus*).

Ditylenchus destructor Thorne (potato rot nematode) is a migratory endoparasite, mainly affecting the underground parts of plants. This nematode has been isolated from rotted roots exhibiting severely stunted growth from two commercial hop yards in New Zealand. Stunted plants and neighboring healthy plants had 310–520 and 0–20 nematodes per 100 ml of soil, respectively. Although no biological races of *D. destructor* have been characterized, isolates from hop, iris, and mint differed in host range, suggesting such races do occur. *D. humuli* has been associated with a severe rot of hop roots in Ukraine and was widespread in Russian yards. However, *D. humuli* has not yet been accepted widely as a valid species.

The ectoparasitic nematode *Tylenchorhynchus dubius* Bütschli has also been reported to cause damage to the main root system of hop plants in Ukrainian hop yards. Severe nematode damage was observed in a commercial hop yard in Belgium where *Heterodera humuli* Filipjev, *Pratylenchus penetrans* (Cobb) Filipjev & Schuurmans Stekhoven, *T. dubius*, and a *Helicotylenchus* sp. were present. Application of the nematicide oxamyl resulted in the recovery of plants and normal yields, and treatment was considered to be economically justified. *Pratylenchus crenatus* Loof and *P. penetrans* have also been associated with hop in Portugal, and the latter species has further been associated with hop in New Zealand. *Meloidogyne* spp., *P. crenatus*, and *Helicotylenchus dihystera* (Cobb) Sher were recorded in hop yards in Tasmania, Australia, at generally low numbers.

Many species of ectoparasitic Longidoridae have been found in German hop yards, including *Longidorus attenuatus* Hooper, *L. caespiticola* Hooper, *L. elongatus* (de Man) Thorne & Swanger, *L. goodeyi* Hooper, *L. intermedius* Kozlowska & Seinhorst, *L. leptocephalus* Hooper, *L. macrosoma* Hooper, *L. profundorum* Hooper, *Paralongidorus maximus* Bütschli, *Xiphinema diversicaudatum* (Micoletzky), and *X. dentatum* Sturhan.

There have been limited reports of root knot nematodes (*Meloidogyne* spp.) on hop. In China, *M. incognita* (Kofoid & White) Chitwood was reported to invade and develop within roots of cultivar Comet, but development was restricted in a resistant locally bred cultivar.

Selected References

Abrantes, I. M. O., Faria, C. A. T., and Santos, M. S. 1987. Root-lesion nematodes (*Pratylenchus* spp.) in Portugal. Nematol. Mediterr. 15:375-378.

Foot, M. A., and Wood, F. H. 1982. Potato rot nematode, *Ditylenchus destructor* (Nematoda: Tylenchida), infecting hops in New Zealand. New Zeal. J. Exp. Agric. 10:443-446.

Katalan-Gateva, S., and Konstantinova-Milkova, M. 1973. Two species of parasitic nematode on hop in Bulgaria. Part 1. Godishnik na Sofiiskiya Universitet Biologicheski Fakultet 68:49-56.

McNamara, D. G., and Eppler, A. 1989. The Longidoridae occurring in the German hop growing regions. Pages 195-197 in: Proc. Int. Workshop Hop Virus Dis. Giessen. Eugen Ulmer GmbH & Co., Stuttgart, Germany.

Mikhailyukov, U. S., and Sigareva, D. D. 1981. The nematode fauna of hops in the Zhitomir region. Vestn. Zool. 2:41-46.

Pelerents, C., Heungens, A., Dejonckheere, W., and Barbez, D. 1983. Control of nematodes in hops. Med. Fac. Landbouww. Rijksuniv. Gent. 48:75-77.

Romascu, E., and Macovei, A. 1989. The effect and structure of nematode communities in some hop fields. Buletinul de Protectia Plantelor 2:45-54.

Zhang, H., and Zheng-Li, L. 1989. Histological studies on the root-knot nematodes nonresistant and resistant root apex of *Humulus lupulus* L. Acta Bot. Sin. 31:576-580.

(Prepared by F. S. Hay and S. J. Pethybridge)

Diseases Caused by Bacteria and a Phytoplasma

Several bacterial diseases of hop have been reported, but these diseases appear to be of minor economic importance. It is interesting to note that few bacterial diseases are important on this crop even though climatic conditions in many moist, maritime regions of hop production are generally conducive to bacterial diseases on numerous other hosts. Accurate diagnosis of bacterial diseases on hop and other plants requires isolation of the causal agent(s) and confirmation of the organism by laboratory assays and pathogenicity studies.

Crinkle Disease

Crinkle disease of hop was first described in Europe in 1930, and subsequent reports of the disease appear in literature published in the 1960s and 1970s. The disease appears to be of little importance in most regions of hop production, although crinkle disease was listed among the most important diseases of hop in Poland in the 1960s. Yield losses of 80% have been attributed to crinkle disease in Europe.

Symptoms

Two distinct symptoms have been attributed to crinkle disease: a line pattern mosaic and chlorotic striping. The intensity of the line pattern mosaic symptoms was reported to vary between years, and symptoms included light spotting and a continuous mosaic pattern that later developed into stripes and rings that sometimes became necrotic. These disease symptoms were reproduced on healthy hop plants after about three months of incubation following mechanical inoculation with leaf homogenates. Softwood cuttings taken from diseased plants rooted poorly. The line pattern mosaic symptoms appear indistinguishable from those caused by *Apple mosaic virus*. Additionally, a serotype of *Prunus necrotic ringspot virus* (presumably *Apple mosaic virus*) has been reported to cause necrotic ringspot symptoms on plants affected by crinkle disease in Germany. Therefore, the line pattern mosaic symptom attributed to crinkle disease appears to be caused by *Apple mosaic virus*.

The striping symptoms appear as yellowish-green stripes between the main leaf veins. Striping symptoms can be reproduced on cucumber cotyledons by mechanical inoculation and on hop, hemp, and nettle by grafting onto diseased shoots. Cones of affected plants have proliferated growth and develop into a phyllody. Similar symptoms can be induced on hemp flowers by grafting onto diseased hop plants.

Causal Organism

Early literature regarding the causal agent of crinkle disease is contradictory and confusing. The disease has been attributed to a virus or virus complex, zinc deficiency, and a phytoplasma. Further complicating the etiology of the disease is that some symptoms attributed to crinkle disease are highly similar to or were confused with those caused by an ilarvirus. However, there is no evidence that a virus or virus-like pathogen is the causal agent of crinkle disease.

A fastidious bacterium-like organism originally described as a rickettsia-like organism is now thought to cause crinkle disease, although definitive evidence for the causal agent is lacking because the presumptive pathogen has not been grown in axenic culture, precluding the completion of Koch's postulates. The role of a fastidious bacterium-like organism in crinkle disease is supported by direct observation of pleomorphic cells in the protoplasm of roots, petioles, and midribs of leaves of diseased but not healthy hop plants. Further evidence is that nearly complete remission of symptoms can be achieved by treatment of diseased plants with penicillin or tetracycline. Successful transmission of the presumptive pathogen by means of hop leaf homogenates to cucumber indicator plants results in the development of crinkle-like disease symptoms on inoculated plants. Pleomorphic bacteria-like particles have been observed in inoculated cucumber plants but not noninoculated plants.

In electron micrographs, cells of the presumptive pathogen are circular to oval and range in length from 1.0 to 1.8 µm, with diameters ranging from 150 to 800 nm. Cells possess a double membrane that surrounds protoplasmic material, which appears dense and granular. Division of the bacterium-like organism by binary fission is commonly observed. In parenchyma cells of diseased plants, small (0.2–0.3 µm), imperceptible bodies, stainable with Giemsa stain, are apparent through the use of light microscopy. Such bodies are occasionally observable in phloem cells of roots and stems of diseased plants, but the organism tends to be most prevalent in xylem tissues.

Disease Cycle and Epidemiology

Very little is known about the disease cycle and epidemiology of crinkle disease. The pathogen is known to be transmitted through grafting and softwood cuttings taken from diseased plants. A leafhopper vector of the crinkle disease pathogen has been speculated, which would be consistent with other plant diseases caused by fastidious bacteria-like organisms.

Management

In general, management of plant diseases caused by fastidious bacteria-like organisms is similar to preventative measures employed for diseases caused by viruses. Softwood cuttings and rhizomes should be taken only from hop plants free of crinkle disease. Remission of crinkle symptoms can be achieved by fertilization with zinc salts applied soon after disease symptoms appear, apparently because translocation of zinc is inhibited in affected plants.

Selected References

Chod, J., Polák, J., Novák, M., and Kříž, J. 1974 Vorkommen von mykoplasmaähnlichen Organismen in Hopfenblättern mit nekrotischem Kräuselmosaik. Phytopathol. Z. 80:54-59

Schmidt, H. E., Wrazidlo, W., Bergmann, W., and Schmelzer, K. 1972. Nachweis von Zinkmangel als Ursache der Kräuselkrankheit des Hopfens. Biol. Zentralbl. 91:729-742.

Vanek, G., Ulrychová, M., Blattný, C., Jokeš, M., and Novák, M. 1976. Rickettsialike structures in hop plants with crinkle disease and therapeutic effect of antibiotics. Phytopathol. Z. 87:224-230.

(Prepared by D. H. Gent and C. T. Bull)

Crown Gall

Crown gall was among the first bacterial diseases to be described on plants. Crown gall occurs worldwide on over 600 diverse dicotyledonous and gymnospermous plant species, including almond, apple, blackberry, cherry, chrysanthemum, daisy, grape, nectarine, peach, pear, plum, raspberry, rose, tomato, sunflower, and walnut among many other common orchard, field, and nursery crops. The host range of some hop isolates of the pathogen is reported to include at least 157 plant species representing 37 families.

Reports of crown gall on hop date back to at least 1929, and the disease has been reported from most countries where hop is or has been grown commercially, including Argentina, Australia (Tasmania), Bulgaria, China, Czechoslovakia, the former Soviet Union, and Yugoslavia. In the United States, crown gall has been reported on hop in California, Oregon, and Washington. Relatively little research has been conducted on crown gall on hop, and the disease seems to be of minor importance on established hop plants. However, reports of severe outbreaks in nurseries and young plantings in South Africa and the Pacific Northwest in the United States, where entire plants were killed, suggest that crown gall can cause economic damage to hop in certain situations.

Symptoms

The most characteristic symptom of crown gall on hop is the presence of soft and fleshy to hard tumors (galls) on bines at or near the soil surface close to the crown (Fig. 98). The galls have a smooth to rough surface texture and vary in size, with diameters ranging from 0.5 to 10.0 cm or larger. In some instances, disorganized root and shoot growth may develop from galls (Fig. 99). A "hairy-root" symptom was noted to occur on the hop cultivar Early Cluster in Oregon. The bines emerging from galls are generally smaller in diameter than the bines below the gall. Occasionally the entire crown becomes covered in galls. Galls may be colonized by secondary organisms, resulting in rot, and affected bines can be girdled completely, leading to wilting. In such cases, no healthy bines may develop and the entire plant may die within 2 years of infection.

Damage to developing bines tends to be most pronounced in hop nurseries and young plants, and the disease may occur on older plants without causing obvious symptoms or appreciable damage. Affected plants may occur randomly throughout a hop

Fig. 98. Characteristic gall on a hop bine caused by *Agrobacterium tumefaciens*. (Courtesy ARC-Plant Protection Research Institute, South Africa)

Fig. 99. Disorganized root growth and crown galls on rhizome caused by *Agrobacterium tumefaciens*. (Courtesy ARC-Plant Protection Research Institute, South Africa)

yard, or in severe cases, a high proportion (>80%) of plants may become diseased, as has been reported in South Africa and the Pacific Northwest in the United States.

Causal Organism

Crown gall of hop is caused by the bacterium *Agrobacterium tumefaciens* (E. F. Smith & Townsend) Conn. Three biotypes of *A. tumefaciens* were originally described (biotypes I, II, and III; biovars 1, 2, and 3) based on laboratory assays. Biovars 1, 2, and 3 are now referred to as *A. tumefaciens*, *A. rhizogenes* (Riker et al.) Conn, and *A. vitis* Ophel and Kerr, respectively. Although both *A. tumefaciens* biovar 1 and *A. radiobacter* (Beijerinck and van Delden) Conn biovar 1 were reported to cause crown gall on hop, the disease is currently attributed to *A. tumefaciens*.

A. tumefaciens is a gram-negative, non-spore-forming, rod-shaped bacterium that possesses one to six peritrichous flagella. Cells are 0.6–1.0 µm wide and 1.5–3 µm long. In axenic culture, the bacterium grows optimally at temperatures of 25–28°C and pH 6–7. Colonies are circular, convex, smooth, and nonpigmented to beige. Copious amounts of extracellular polysaccharide are produced on culture media containing carbohydrates. Biochemical tests can be used to differentiate *A. tumefaciens* from *A. rhizogenes* and *A. vitis*, including assays for determining motility, production of 3-ketoglycosides, pectolytic activity at pH 4.5, and acid-clearing on media containing yeast extract, dextrose, and calcium carbonate. Tumorigenic and nontumorigenic strains of *A. tumefaciens* exist, but biochemical tests do not differentiate among these strains. *Agrobacterium* species can also be differentiated by fatty acid profiling, metabolic fingerprinting, serological methods, and PCR assays.

Pathogencity of *A. tumefaciens* is associated with the presence of a large tumor-inducing plasmid, which contains virulence genes and a transfer DNA (T-DNA) region that includes genes for biosynthesis of plant hormones and unusual amino acid derivatives (opines). The virulence genes mediate transfer and integration of the T-DNA into plant cells, leading to a proliferation of cell division and growth (e.g., the crown galls). Infected cells also produce opines that are utilized by the pathogen. After the T-DNA has integrated into the host genome, the proliferated cell growth becomes an autocatalytic process that does not depend on the presence of the pathogen. The unique interaction allowing DNA to be transferred from *A. tumefaciens* and integrated into a host's genome is the foundation for genetic engineering of plants, and the bacterium has been utilized extensively in molecular biology to transform numerous plant species and fungi with native and foreign DNA.

Disease Cycle and Epidemiology

The disease cycle and epidemiology of crown gall on hop has not been investigated extensively, but it is presumably similar to that on other crops. The pathogen is a common inhabitant of soil and the rhizosphere of many dicotyledonous plants. Soil populations of the bacterium are thought to originate from gall residue and other infested plant residue. Bacterial cells are readily disseminated within and among fields in infected plant materials, irrigation water, infested soil, and on cutting tools and equipment. The bacterium survives in association with infested soil, crop debris, and seed and from within diseased plants. The presence and survival of *A. tumefaciens* in soil in hop yards is unknown, although the bacterium has been documented to persist in soil for as long as 2 years in the absence of a host plant and is commonly isolated from soil and rhizosphere samples. The most severe outbreaks of crown gall appear to result from propagation of infected plant material.

Infection occurs through plant wounds created by frost injury or mechanical damage caused by pruning or other cultural practices. The presence of plant phenolic compounds associated with wounds activates virulence genes in the pathogen and initiates the infection process. Galls can be produced on susceptible cultivars such as Fuggle and Late Cluster within 30 days of inoculation with the pathogen. However, plants may be infected without causing appreciable damage unless entire crowns become diseased and no or few healthy shoots are produced.

Management

Management of crown gall is most important in nurseries and during establishment of new hop yards. The pathogen is readily disseminated in infected plant material, which may or may not have symptoms of disease. Therefore, softwood cuttings and rhizomes should be harvested only from plants free of the crown gall bacterium. Severely infected plants should be removed from nurseries and established hop yards and destroyed.

Populations of the bacterium decrease over time in the absence of a suitable host, although detectable levels of the pathogen may persist in soil for 2 years or longer. In situations where severe crown gall outbreaks have occurred, a 2- to 3-year rotation to a monocotyledonous crop is advisable.

Biological control of crown gall on aerial stems of greenhouse-grown hop plants has been achieved with avirulent strains of *A. tumefaciens* (originally called *A. radiobacter* K84 and D286), which produce bacteriocins (antibacterial compounds) that suppress some strains of *A. tumefaciens*. Formulations of biocontrol strains are available commercially. Strains of *A. tumefaciens* insensitive to bacteriocins produced by the biological control strains may also occur, and the efficacy of biological control is reduced where insensitive strains of *A. tumefaciens* are prevalent. Biological control of crown gall is less effective on plants with latent infections of *A. tumefaciens*.

Host resistance to crown gall has been reported in other crops but has not been investigated thoroughly with hop. Results of an evaluation of 40 hop cultivars published in 1945 listed some cultivars that did not develop galls following artificial inoculation, although the small sample sizes used in that study preclude strong recommendations on selection of cultivars resistant to crown gall.

Selected References

Chen, X.-Y., and Xiang, W.-N. 1987. Studies on *Agrobacterium tumefaciens* isolated from hops in China. Acta Microbiol. Sin. 27:10-16.

Haas, J. H., Moore, L. W., Ream, W., and Manulis, S. 1995. Universal primers for detection of pathogenic *Agrobacterium* strains. Appl. Environ. Microbiol. 6:2879-2884.

Hoerner, G. R. 1945. Crown gall of hops. Plant Dis. Rep. 29:98-110.

Horlemann, C., Schwekendiek, A., Höhnle, M., and Weber, G. 2003. Regeneration and *Agrobacterium*-mediated transformation of hop (*Humulus lupulus* L.). Plant Cell Rep. 22:210-217.

Sawada, H., Ieki, H., and Matsuda, I. 1995. PCR detection of Ti and Ri plasmids from phytopathogenic *Agrobacterium* strains. Appl. Environ. Microbiol. 61:828-831.

Serfontein, S., and Staphorst, J. L. 1993. Crown gall of hop caused by *Agrobacterium tumefaciens* biovar 1 in South Africa. Plant Pathol. 43:1028-1030.

Smith, E. F., and Townsend, C. O. 1907. A plant-tumor of bacterial origin. Science 25:671-673.

Young, J. M., Kerr, A., and Sawada, H. 2005. Genus II. Agrobacterium. Pages 341-345 in: Bergey's Manual of Systematic Bacteriology (2nd ed.) Volume Two, The Proteobacteria. D. J. Brenner, N. R. Krieg, and J. T. Staley, eds. Springer, New York.

(Prepared by D. H. Gent and C. T. Bull)

Bacterial Diseases of Minor Importance

Bacterial Blight

Reports of *Pseudomonas cannabina* (ex Šutić and Dowson) Gardan et al. as a pathogen of hop can be found in the popular press, although the limited information on this organism in the scientific literature indicates the host range of *P. cannabina* does not include hop. Recent attempts to generate bacterial blight symptoms on hop by artificial inoculation with *P. cannabina* have also failed. On hemp, *P. cannabina* has been described in Bulgaria, Germany, Hungary, Italy, Romania, Yugoslavia, and the former Soviet Union.

Xanthomonas Leaf Spot

Xanthomonas leaf spot is characterized by small water-soaked spots on leaves that later coalesce into large lesions and become necrotic. A similar disease on hemp has been reported in Japan, Korea, and Romania and was attributed to *Xanthomonas campestris* pv. *cannabis* Severin. The host range of the pathogen is reported to include common bean, cucumber, mulberry, soybean, tobacco, and a *Pelargonium* sp., and reports of this bacterium as a pathogen of hop can be found in the popular press. The bacterium is able to infect hop by artificial inoculation, but outbreaks of Xanthomonas leaf spot on hop in the field have not been reported. Thus, naming this organism as a natural pathogen of hop is questionable.

Diseases of Unknown Etiology

A *Corynebacterium* sp. has been implicated in a poorly understood disease of hop in Japan, named simply "bacterial disease." Disease symptoms attributed to the bacterium include (i) a proliferation of crown buds ("cauliflower-like" symptom), (ii) an abundance of thin, abnormally shrunken shoots that fail to develop more than 30 cm ("witch's-broom" symptom), (iii) degeneration of leaves from well-developed (1.5–2 m high) bines so the bine appears as a slender whip ("whip-shaped" symptom), and (iv) a rapid death of the main bine that releases apical dominance of lateral branches that later develop normally ("dry tip-rot" symptom). Symptoms associated with bacterial disease are very similar to those of "small hop disease" described in the United Kingdom in 1932 and hop shoot proliferation associated with a phytoplasma. Stow and Ihara (1955) proposed the name *Corynebacterium humuli* Stow & Ihara for the causal agent of bacterial disease, but the etiology of this disease is not understood completely. A supplementary note by the authors indicated the presumptive *Corynebacterium* pathogen rarely induced disease symptoms by artificial inoculation, and the authors suggested that another organism was likely involved in disease development. Following the initial reports, *Corynebacterium* sp. has been reclassified into several new genera and the name *C. humuli* has been lost. A coryneform bacterium was not recognized in the List of Prokaryotic Names with Standing in Nomenclature produced by the International Journal of Systematic Bacteriology because type cultures were not preserved or available.

Selected References

Euzéby, J. P. 1997. List of Bacterial Names with Standing in Nomenclature: A folder available on the Internet. Int. J Syst. Bacteriol. 47:590-592.

Severin, V. 1978. Ein neues pathogenes Bakterium an Hanf—*Xanthomonas campestris* pathovar *cannabis*. Arch. Phytopathol. Pflanzenschutz 14:7-15.

Stow, I. 1954. A new disease of hop due to a bacterium. Proc. Jpn. Acad. 30:226-231.

Stow, I., and Ihara, K. 1955. On a new disease-germ of the hop plant. Proc. Jpn. Acad. 31:294-299.

Stow, I., Hamaguchi, T., Ihara, K., Sakai, M., Sato, K., Uehara, Y., and Mori, Y. 1955. A new disease of hop due to a bacterium (A preliminary report). Bull. Brew. Sci. 1:3-6.

Šutić, D., and Dowson, W. J. 1959. An investigation of a serious disease of hemp (*Cannabis sativa* L.) in Jugoslavia. Phytopathol. Z. 34:307-314.

(Prepared by D. H. Gent and C. T. Bull)

Hop Shoot Proliferation

Hop shoot proliferation has been described in Poland and is associated with phytoplasma infection. The disease occurs rarely and seems to be of little economic concern in most regions of hop production. Phytoplasmas are plant pathogens associated with diseases in several hundred plant species including many important food, vegetable, and fruit crops, ornamental plants, and timber and shade trees.

Symptoms

To date, the symptoms of hop shoot proliferation have been observed and described only on cultivars Magnum and Marynka. Pronounced symptoms of phytoplasma infection are manifested on buds and shoots when rhizomes are dug. Affected plants produce numerous weak shoots (Fig. 100). Leaves are small, distorted, and chlorotic. The most severely affected plants are stunted and do not produce flowers or produce only single malformed flowers. Later in the season, shoot proliferation, leaf necrosis and malformation, and premature plant senescence develop, leading to eventual plant death. The severity of symptoms varies among plants, and some of the affected plants may remain asymptomatic. Symptoms of hop shoot proliferation appear similar to those attributed to a *Corynebacterium* sp. in Japan in 1954 and "small hop disease" described in the United Kingdom in 1932.

Causal Organism

Hop shoot proliferation is thought to be caused by aster yellows phytoplasma, group 16SrI-B, now reclassified as '*Candidatus* Phytoplasma asteris', which is associated with symptomatic plants. Phytoplasmas are phloem-limited pleiomorphic bacteria lacking cell walls that inhabit phloem sieve elements in infected plants. Biochemical and physiological characters used for identifying other prokaryotes have not been useful because phytoplasmas have not been grown in axenic culture. In single cross sections, phytoplasmas appear as rounded, pleomorphic bodies with diameters ranging from 200 to 800 µm, although a filamentous morphology has been reported in some hosts. Detection of phytoplasmas is typically accomplished by PCR assays with phytoplasma universal or group-specific primer pairs derived from conserved 16S rRNA gene sequences. The phytoplasmas detected in hop can be identified by direct sequence analysis or restriction fragment length polymorphisms of PCR-amplified rDNA. Serological methods such as ELISA have also been developed for detection of phytoplasmas, but serological

Fig. 100. Shoot proliferation symptoms on hop plants infected with a phytoplasma. Shoot of healthy plant at right. (Courtesy M. Kamińska)

methods are typically of minor importance for phytoplasma detection and identification.

Disease Cycle and Epidemiology

Very little is known about the disease cycle and epidemiology of shoot proliferation disease on hop, although the disease cycle and epidemiology of diseases caused by phytoplasmas on other crops have been investigated extensively. Phytoplasmas are disseminated by phloem-feeding insect vectors, primarily certain species of leafhoppers (family Cicadellidae) and planthoppers (superfamily Fulgoroidea), although psyllids (family Psyllidae) are known to vector certain phytoplasma strains. Numerous polyphagous leafhoppers are able to transmit phytoplasmas in subgroup 16SrI-B, although the identification and specificity of the vectors that transmit the phytoplasma that infects hop are unknown.

Plants and insects can be infected by phytoplasmas, and the specificity of the pathogen in both hosts varies among phytoplasma strains. Insect vectors passively acquire phytoplasmas within minutes or hours of feeding on the phloem of infected plants. Successful transmission of phytoplasmas to plants requires that specific cells in the insect salivary glands be invaded and accumulate high titers of the pathogen. Phytoplasmas survive in insect vectors and infected weeds and crops and may be disseminated in softwood cuttings and rhizomes taken from infected plants. Dispersal of the pathogen over a range of hundreds of kilometers in vectors is known to occur with phytoplasma diseases of other hosts.

Management

Control measures for hop shoot proliferation have not been reported, but the disease seems to occur rarely and its low prevalence and incidence dictate that control measures are probably not needed in most instances. In many other vegetatively propagated crops, management of diseases caused by phytoplasmas is generally accomplished by controlling the leafhopper vector with well-timed insecticide applications, host plant resistance, and selecting pathogen-free plant materials for propagation. Phytoplasmas are sensitive to high temperatures and can be eliminated from infected plants through meristem culture and heat therapy.

Selected References

Salmon, E. S., and Ware, W. M. 1932. The small hop disease. J. S. E. Agric. Coll. Wye 30:22-26.

Solarska, E., Kamińska, M., and Śliwa, H. 2004. First report of phytoplasma infection in hop plants. Plant Dis. 88:908.

Stow, I. 1954. A new disease of hop due to a bacterium. Proc. Jpn. Acad. 30:226-231.

(Prepared by E. Solarska, H. Śliwa, M. Grudzińska-Sterno, M. Kamińska, and D. H. Gent)

Part II. Arthropod Pests

California Prionus Beetle

Prionus californicus Mots. (Cerambycidae: Prioninae), the California prionus, is broadly distributed in western North America where it is an economically important pest of many agricultural and ornamental plants, including apple, rose, lilac, and oak. This species has been a serious pest of hop in the Pacific Northwest for over 60 years.

Biology and Life History

Adult California prionus emerge from pupation sites in the soil from late June through July in the Pacific Northwest of the United States. Adults range in body length from 25–55 mm (females are larger than males) (Fig. 101) and are among the largest cerambycids in North America. Adults are crepuscular, do not feed, and live for only 2–3 weeks, during which time they must mate and lay eggs. Mate location is mediated by female-produced volatile pheromones, while mate recognition is mediated by nonvolatile contact pheromones on the female cuticle. Eggs are laid on or in the soil near the base of plants. A single female can lay 150–200 eggs. Cream-colored larvae emerge from eggs and begin feeding on plant roots and grow from less than 5 mm to as much as 75 mm over a 3- to 5-year period (Fig. 102). Larvae can be found in roots 50–800 mm below the soil surface. Small larvae (less than 15 mm) are found deeper in the root zone than larger larvae and tend to tunnel in roots. Large larvae tend to graze and prune large roots closer to the soil surface. Mature larvae pupate in cells constructed from soil and lined with root material (Fig. 103).

Fig. 101. Female (left) and male (right) California prionus adults. Male is about 40 mm long. (Courtesy J. D. Barbour)

Fig. 102. California prionus larvae. Small larva on left is about 8 mm long; largest larva is about 60 mm. (Courtesy J. D. Barbour)

Damage

Adult California prionus are not known to damage hop plants. However, larvae feeding on the roots of hop plants can cause serious economic losses. Larval feeding damages plant roots, resulting in decreased nutrient uptake, water stress, and reduced plant growth. Severe infestations can completely destroy plant

Fig. 103. California prionus pupa case (top) and pupa (bottom). (Courtesy J. D. Barbour [top] and Ken Gray Image courtesy Oregon State University [bottom])

crowns, resulting in plant death (Fig. 104). Less severe infestations can result in the wilting, yellowing, and death of one or more bines of infested plants (Fig. 105).

Management

Current management of California prionus in hop is deficient and consists only of identifying and removing infested plants. It is necessary to dig up and remove all plants in severely infested yards. Once the infested plants have been removed, the yard can be fumigated or planted to a nonhost crop for 2–3 years to reduce populations before replanting the yard with hop. In this case, there is limited opportunity to produce alternative crops in the interim because the trellis system upon which hop plants are grown must remain in place. There are no hop cultivars or accessions identified as being resistant to California prionus. No economically significant natural enemies have been identified. There are no pesticides registered for control of California prionus in hop. Management through mating disruption or adult trapping techniques may be possible using the volatile pheromone mediating mate recognition in this species.

Selected References

Barbour, J. D., Lacey, E. S., Millar, J. G., and Hanks, L. M. 2007. Cuticular hydrocarbons mediate mating in the Prionine Cerambycid *Prionus californicus* Mots. Ann. Entomol. Soc. Am. 100:333-338.

Bishop, G. W., Blackmer, J. L., and Baird, C. R. 1984. Observations on the biology of *Prionus californicus* Mots. on hops, *Humulus lupulus* L., in Idaho. J. Entomol. Soc. B. C. 81:20-24.

Cervantes, D. E., Hanks, L. M., Lacey, E. S., and Barbour, J. D. 2006. First documentation of a volatile sex pheromone in longhorned beetles (Coleoptera:Cerambycidae) of the primitive subfamily Prioninae. Ann. Entomol. Soc. Am. 99:718-722.

Linsley, E. G. 1959. The ecology of the Cerambycidae. Annu. Rev. Entomol. 4:99-138.

Linsley, E. G. 1962. The Cerambycidae of North America. Part II. Taxonomy and classification of the Parandrinae, Prioninae, Spondylinae and Aseminae. University of California Press, Berkeley.

Linsley, E. G., and Chemsak, J. A. 1997. The Cerambycidae of North America. Part VIII. Bibliography, index, and host plant index. Univ. Calif. Publ. Entomol. 117:1-534.

(Prepared by J. D. Barbour)

Fig. 104. Large California prionus larva in debris of severely damaged hop crown. (Courtesy J. D. Barbour)

Fig. 105. Wilting of plants with severe damage caused by California prionus larvae feeding on hop roots. (Courtesy J. D. Barbour)

Damson-Hop Aphid

The damson-hop aphid, *Phorodon humuli* (Schrank), is one of the dominant pest species in hop-growing regions of the Northern Hemisphere. Because the aphid can cause serious yield losses up to the complete destruction of a crop, *P. humuli* is of major economic importance in hop production where the pest occurs.

Appearance

On hop, the summer host, the adult *P. humuli* is a very small aphid with a body length of 1.2–1.8 mm and an average weight of 360 µg. On *Prunus* spp., the winter host, the aphids are larger at 2.1–2.8 mm long. Their color ranges from pale green to bottle green, with dark green longitudinal stripes on the dorsum of the abdomen. The nymph stages are almost white (Fig. 106). The two siphunculi at the end of the abdomen are pale, cylindrical, and slightly curved at their tips. The most characteristic features that distinguish *P. humuli* from similar aphids are a pair of pointed projections between the antennae on the "forehead" and a triangular projection on the inner side of the first anten-

Fig. 106. Damson-hop aphid colony (top) and closeup of adults (bottom right) and nymphs (bottom center and left). (Courtesy W. F. Mahaffee [top], T. Neill [bottom left and center], and Ken Gray Image courtesy Oregon State University [bottom right])

nal segment (Fig. 106). Winged migrants have a body length of 1.8–2.1 mm and dark brown head, thorax, and antennae (Fig. 107). The projections on the forehead and antennae are much less developed in the winged form.

Biology and Life History

A member of the family Aphididae, *P. humuli* is a holocyclic aphid that overwinters in the egg stage on various *Prunus* spp. The most important winter host plants are sloe (*Prunus spinosa* L.), plum (*P. domestica* L.), cherry plum (*P. cerasifera* Ehrh.), and damson (*P. insititia* L.). In early spring between late February and April, a wingless female (fundatrix) hatches from each egg, feeds on *Prunus* leaves, and asexually produces an average of four generations of wingless female aphids (fundatrigenia). After one to two generations, the fundatrigenia begin to produce winged females (alatae) with the percentage of alatae produced increasing in each subsequent generation. Depending on weather conditions, these winged forms begin to appear in May and start their migration to hop, the only genuine summer host, when a flight threshold temperature (i.e., a temperature minimum during daytime of 13–14°C) is exceeded. In contrast to many other aphid species, migrating *P. humuli* do not have an obligatory long-distance flight phase but are flexible in their migration strategy from winter to summer hosts. *P. humuli* may migrate a few dozen meters or several kilometers by passive stratiform drift. The usual distance is 100–1,000 m, if winter and summer hosts are located that close.

Soon after their arrival on hop, the migrants start producing wingless, asexual female aphids (apterae). Depending on weather conditions and host susceptibility, more than 10 overlapping generations of apterae can develop over the summer, with each generation living 2–4 weeks and producing approximately 30–50 new nymphs. Hence, large populations of *P. humuli* may build up if left uncontrolled.

Contrary to some older reports, *P. humuli* does not produce winged alienicolous forms that move to other hop plants during the summer. Toward the end of August, decreasing day length causes apterae on hop to switch to producing female migrants (gynoparae) that migrate back to the *Prunus* winter host. There, the gynoparae give birth to wingless females (oviparae). Approximately 2 weeks after producing gynoparae, the apterae on hop begin to produce winged males that also migrate to the winter host. On *Prunus* spp. the males locate oviparae by means of a pheromone released from these sexual females and mate. The fertilized oviparae then lay overwintering eggs on or near the buds of the winter host plant. The fall aphid migration usually continues into October or November until it is terminated by the first foliage-killing frost.

Fig. 107. Winged female (alata) of the damson-hop aphid. (Ken Gray Image courtesy Oregon State University)

Damage

P. humuli feeds by inserting its stylet into the phloem strands of the leaf veins to suck plant sap. Heavy aphid infestations may reduce plant vigor and induce defoliation. However, the main problem is infestation of developing cones. Late in the season, *P. humuli* frequently feeds within the cones protected from insecticides, and infested cones soon turn brownish and limp. Hence, even light infestation of cones can significantly reduce their appearance and economic value, and heavy infestation may result in the complete destruction of a crop. In addition, aphids consume tremendous amounts of the sieve tube sap, much of which passes through the digestive system more or less unmetabolized to be excreted as a sugary ooze (honeydew) on foliage. The excreted honeydew serves as a medium for rapid growth of sooty mold fungi on leaves and cones, making the latter unmarketable (Fig. 108). Furthermore, *P. humuli* may serve as a vector for the transmission of a number of carlaviruses such as *Hop mosaic virus* and *American hop latent virus*.

Control

Aphid control is usually an essential element of hop production. Uncontrolled populations of *P. humuli* do not always result in total crop loss, and in some years aphid populations are negligible because of the action of natural enemies. Nevertheless, for almost a century, farmers have used chemical control for damson-hop aphid. Early in the twentieth century, farmers used soft soap, quassia, and nicotine for aphid control. These were later replaced with more effective compounds like organophosphates in the 1950s and insecticides of other chemical classes after that. However, because they were used exclusively for long periods, *P. humuli* has developed resistance to many of these compounds, such as carbamates, endosulfan, organophosphates, pyrethroids, and, most recently, imidacloprid. Thus, effective use of chemical insecticides has to include proper rotation of insecticide classes so that no class is applied more than once in a season. In addition, all insecticides should be applied at the recommended rates and volumes.

Fig. 108. Aphid damage and black sooty mold on hop leaves and cones. (From left to right courtesy W. F. Mahaffee, D. H. Gent, W. F. Mahaffee, and M. E. Nelson)

Biological control of *P. humuli* by natural enemies has been documented for almost a century. The range of naturally occurring predators includes various ladybirds, anthocorids, lacewing and hoverfly larvae, and parasitoids such as aphidiid wasps. Together, such natural enemies can provide satisfactory control of aphid populations, provided that the environmental conditions are more favorable for the natural enemies than the aphids. There have been numerous trials involving the release of mass-reared antagonists to control *P. humuli*. However, most such attempts have failed, since it is impossible to artificially generate a predator density comparable to the numbers achieved under optimal natural conditions in the yard. A more promising approach to augmenting naturally occurring predators and parasites is provided by recent experiments in which synthetic synomones or kairomones that act as attractants for beneficial insects were used. Other chemoecological research targets *P. humuli* directly by the use of compounds that may act as repellents or as attractants for the aphids. The sex pheromone that attracts male *P. humuli* to the oviparae has been identified, but it is of little practical use for control. Under specific temperature and relative humidity conditions, entomopathogenic fungi can parasitize *P. humuli* and significantly reduce populations within a few days.

The keystone to an integrated approach to the control of *P. humuli* is the use of hop genotypes that are resistant or tolerant to this pest and acceptable to the brewing industry. At present, few commercial cultivars possess these traits. Because of the difficulties in developing aphid-resistant or tolerant cultivars, *P. humuli* management will continue to be dependent on the integration of many methods for managing natural antagonists combined with the judicious use of aphicides.

Selected References

Campbell, C. A. M., and Muir, R. C. 2005. Flight activity of the damson-hop aphid, *Phorodon humuli*. Ann. Appl. Biol. 147:109-118.

Campbell, C. A. M., Petterson, J., Pickett, J. A., Wadhams, L. J., and Woodcock, C. M. 1993. Spring migration of damson-hop aphid, *Phorodon humuli* (Homoptera, Aphidiidae), and summer host plant-derived semiochemicals released on feeding. J. Chem. Ecol. 19:1569-1576.

Darby, P., and Campbell, C. A. M. 1996. Aphid-resistant hops—the key to integrated pest management in hops. Pages 893-898 in: Proc. Brighton Crop Prot. Conf.-Pest Dis., British Crop Protection Council, Hampshire, U.K.

Hardie, J., and Campbell, C. A. M. 1998. The flight behaviour of spring and autumn forms of the damson-hop aphid, *Phorodon humuli*, in the laboratory. Pages 205-212 in: Aphids in Natural and Managed Ecosystems. J. M. Nieto Nafría and A. F. G. Dixon, eds. Universidad de León (Secretariado de Publicaciones), León, Spain.

Lösel, P. M., Lindemann, M., Scherkenbeck, J., Campbell, C. A. M., Hardie, J., Pickett, J. A., and Wadhams, L. J. 1996. Effect of primary-host kairomones on the attractiveness of the hop-aphid sex pheromone to *Phorodon humuli* males and gynoparae. Entomol. Exp. Appl. 80:79-82.

Taimr, L., and Křiž, J. 1978. Stratiform drift of the hop aphid (*Phorodon humuli* Schrank). Z. Angew. Entomol. 86:71-79.

Weihrauch, F., and Moreth, L. 2005. Behavior and population development of *Phorodon humuli* (Schrank) (Homoptera: Aphididae) on two hop cultivars of different susceptibility. J. Insect Behav. 18:693-705.

Wright, L. C., Cone, W. W., and James, D. G. 2005. Sources of spring and fall hop aphid, *Phorodon humuli* (Schrank) (Homoptera: Aphididae) migrants in south central Washington. J. Entomol. Soc. B. C. 102:21-26.

(Prepared by F. Weihrauch)

Garden Symphylan

Garden symphylans (*Scutigerella immaculata* Newport) are small, white, centipede-like soil arthropods (Fig. 109) that are generalist root-feeding pests. Garden symphylans have been reported as a pest of hop only in the maritime climate of the Pacific Northwest of the United States.

Biology and Life History

Garden symphylans are not insects, but members of the class Symphyla. They measure about 0.6 cm long when mature and have six to 12 pairs of legs (depending on age), which make symphylans easy to differentiate from common soil insects that have only three pairs of legs. Eggs, adults, and immature garden symphylans can be found together throughout most of the year. In the maritime Pacific Northwest, garden symphylans are common in many alluvial soils with high organic matter. Temperature plays a key role in regulating oviposition, with the greatest number of eggs most commonly deposited in the spring and fall. Garden symphylans are quite active and mobile for their size, darting rapidly and moving vertically and horizontally through the soil profile by means of soil pores and channels made by roots and other soil organisms. Garden symphylans tend to aggregate in the top 15 cm of soil when the soil is moist and warm and move to deeper soil strata when soils become very dry or cool. Surface populations may be significant whenever temperatures are fairly warm, sufficient moisture is present, and roots are shallow or absent; the greatest populations occur in May and June. When conditons are such that symphylans are at the soil surface, they are often localized in patches that can be identified by sampling before and after planting.

Damage

Garden symphylans are extremely common in many soils where they do not cause economic damage. However, symphylans are often implicated in crop damage when relatively high populations (i.e., greater than 1 per 6,000 cm^3 of soil) are found concentrated in close proximity to stunted plants damaged by feeding on fine roots. Root feeding by garden symphylans can lead to mild to severe stunting, reduced plant vigor, and mortality (Fig. 110). Feeding may lead to poor plant establishment and may contribute to the decline of established hop yards. Root damage may also render plants more susceptible to some soil-borne pathogens.

Management

Garden symphylans are particularly difficult to manage in hop in part because of the inability to cultivate near the perennial crowns, absence of effective and persistent soil-applied insecticides, and difficulties associated with delivering insecticides through the soil profile to where symphylans occur. No simple, inexpensive, and completely reliable method for controlling garden symphylans has been developed.

Fig. 109. Garden symphylan adult. (Ken Gray Image courtesy Oregon State University)

Fig. 110. Diminished plant development in a 5-year-old yard of cultivar Willamette in Oregon with a severe infestation of garden symphylan. (Courtesy W. F. Mahaffee)

In many cases, well-established plants can tolerate a moderate amount of damage. However, control is critical during plant establishment and in early spring. Tillage and pesticides are the most common tactics used to manage symphylan populations. In hop yards, tillage can be used immediately before planting and in the spring to crush garden symphylans and disrupt their movement. The effects of tillage may vary with the type of implements used, soil type, and soil moisture at time of tilling. In general, the more disruptive the tillage, the greater the effect it has on garden symphylan movement and feeding. In conjunction with tillage, pesticides are commonly used to manage garden symphylans.

Fumigants and organophosphate and carbamate pesticides have been the most effective compounds for controlling garden symphylans. Plant protection is likely achieved through direct mortality of symphylans as well as repellency from the root zone. Pesticides are most effective if broadcast and incorporated as close to hop plants as possible. Banded/incorporated methods can be used to apply products close to plants but are not as effective as broadcast applications. Sampling can provide information to manage garden symphylans effectively. Sampling is commonly conducted from April to June when garden symphylans are present in the top 15–30 cm of the soil surface.

Selected References

Edwards, C. A. 1958. The ecology of Symphyla. Part I: Populations. Entomol. Exp. Appl. 1:308-319.

Michelbacher, A. E. 1937. Control of the garden centipede in California. J. Econ. Entomol. 30:887-891.

Umble, J. R., and Fisher, J. R. 2003. Sampling considerations for Garden Symphylans (Order:Cephalostigmata) in western Oregon. J. Econ. Entomol. 96:969-974.

(Prepared by J. R. Umble)

Hop Flea-Beetle

The leaf-beetle recognized as the "hop flea-beetle" pertains to two different species of the subfamily Alticinae (Coleoptera, Chrysomelidae). In North America it is represented exclusively by *Psylliodes punctulatus* Melsheimer, whereas in the Palaearctic region the damage on hop can be attributed chiefly to *P. attenuatus* Koch. In a few cases, other Alticinae (such as *Chaetocnema concinna* Marsham and *Phyllotreta* spp.) have been reported as flea-beetles damaging hop. However, those records probably pertained to species feeding on cruciferous or amaranthus cover crop plants in the yards and were likely linked erroneously to hop. Regardless, the vast majority of flea-beetles attacking hop are either *P. punctulatus* or *P. attenuatus*, depending on the continent. Because their biology is very similar, both species are treated as one pest in the following notes.

Appearance

Adult *P. attenuatus* are 2.0–2.8 mm long and 1.0–1.4 mm wide and have a brassy, metallic green to bronze color. *P. punctulatus* is slightly smaller, measuring 1.5–2.5 mm long and about half as wide as *P. attenuatus*, and it is bronzy black in color. Males are generally smaller than females. Both species possess strongly developed hind "thighs" that enable the beetles to jump up to 60 cm, hence the name hop flea-beetle.

Biology and Life History

Hop flea-beetles overwinter in the adult stage. Many types of hibernation shelters are used, such as cracks in wooden hop poles, hollow stubs of hop bines, beneath stones or plant detritus, under bark, and in loose-packed soil. Since hop flea-beetles gradually become active at temperatures above approximately 5°C, the beetles start to leave their shelters early in the spring from early March to late April, depending on weather conditions. By the beginning of May, the population has usually emerged completely from hibernation. Copulation and oviposition occur mainly during May and June. By early July, most adult beetles have died. Oviposition takes place in the upper layer (approximately 0.5–2.0 cm) of moist soil with the help of the female's long ovipositor. In laboratory experiments with *P. punctulatus,* the number of eggs laid per female was four to 18 and the duration of the egg stage was 19–22 days. The egg stage for *P. attenuatus* was 10–12 days. Most larvae hatch from the eggs during June and live in the soil at a depth of up to 25 cm. They feed on the fine filamentous roots of hop and other plants like chickweed without noticeably damaging them. The larval stage lasts approximately 4–5 weeks until larvae pupate in the soil for another 3–5 weeks. From late July to August, the next generation of adult beetles emerges from their pupae to start hibernation again in late autumn. Hence, there is only one generation per year that hibernates and has an adult lifespan of approximately 10 months.

Damage

As soon as the first hop shoots appear above ground in spring, the young bines and leaves may be attacked by hop flea-beetles. During the early growth stages, adult beetles gnaw through the upper cuticle and epidermis of leaves and feed on green parenchymal cells. Typically, this injury gives leaves the appearance of being covered with fine shot holes. In severe cases, many leaves are skeletonized, and photosynthesis may be reduced and growth retarded. There are some reports of extreme cases of hop flea-beetle damage that caused the complete defoliation of young hop plants in entire yards. However, such injury is usually confined to the lower 1–2 m of the bines, and the rapid growth of hop in spring normally allows the plants to outgrow the damage relatively quickly. Later in the season, the newly emerged generation of adult beetles can feed on the bracteoles of young cones. In some cases this damage reportedly may cause the complete destruction of a crop, but this degree of damage is extremely rare.

Control

The hop flea-beetle can be controlled by applications of systemic insecticides used against soil insects or damson-hop aphid. Otherwise, registered insecticides may be applied to the foliage. Normally a single application is sufficient and is most effective if made early in the season when the plants are less than 1 m high.

Selected References

Chittenden, F. H. 1909. The hop flea-beetle (*Psylliodes punctulata* Melsh.). U.S. Dep. Agric. Bur. Entomol. Bull. 66, part VI:71-58, plates V-VII.

Heikertinger, F. 1913. *Psylliodes attenuata* Koch, der Hopfen- oder Hanf-Erdfloh. II. Teil. Morphologie und Bionomie der Imago. Verh. Kaiserl.-Königl. Zool.-Bot. Ges. Wien 63:98-136.

Newton, H. C. F. 1929. Observations on the biology of some flea-beetles of economic importance. 4. On the biology of *Psylliodes attenuata* Koch, the hop flea-beetle. J. S. E. Agric. Coll. Wye 26:154-157.

Parker, W. B. 1910. The life history and control of the hop flea-beetle (*Psylliodes punctulata* Melsh.). U.S. Dep. Agric. Bur. Entomol. Bull. 82, part IV:33-58, plates III-IV.

Tölg, F. 1913. *Psylliodes attenuata* Koch, der Hopfen- oder Hanf-Erdfloh. I. Teil. Morphologie und Biologie der Präimaginalstadien. Verh. Kaiserl.-Königl. Zool.-Bot. Ges. Wien 63:1-25.

Whitehead, P. F. 2000. A recent English record of *Psylliodes attenuata* (Koch, J. D. W., 1803) (Coleoptera: Chrysomelidae) in a riparian situation. Entomol. Gaz. 51:129-131.

(Prepared by F. Weihrauch)

Hop Looper and Other Lepidoptera

Hop looper (*Hypena humuli* (Harris)) is a Noctuid moth native to and widely distributed across the continental United States and Canada. It appears to be restricted to hop and nettles (*Urtica* spp.), which are its only recorded host plants. Hop looper was first reported as a pest of hop in the mid-1800s, and until recently it has generally been regarded as a relatively minor and sporadic problem. However, it now seems to be developing into a more frequent and damaging pest, probably as a result of the gradual shift away from broad-spectrum organophosphate compounds for the routine control of aphids and mites to more selective pesticides that provide no incidental control of loopers.

Biology and Life History

Adult hop loopers are sexually dimorphic; the females have a distinctive W-shaped dark patch along the leading edge of each forewing. This mark is generally obscured in the males by the darker and more uniform color of their wings. The overall wingspan is approximately 26 mm. During the day, the adults typically remain concealed within the hop foliage, flying only when disturbed. Both sexes have the distinctive, elongated "snout" that gave this species its original common name of "hop-vine snout moth."

Fig. 111. Hop looper larva. (Courtesy D. G. James)

The larvae are pale green and have a narrow white line on each side of the dorsum and a slightly flattened appearance (Fig. 111). Larvae have four pairs of prolegs: three pairs on abdominal segments 4, 5, and 6, plus a final pair on the terminal abdominal segment. They move with a characteristic "looping" motion and appear to feed mainly at night. During the day, they are typically found lying motionless along the main vein on the underside of a leaf or stretched out along the petiole. The younger instars respond to disturbance by dropping from the leaves on a silken thread, while the larger larvae may react with a violent side-to-side thrashing motion.

Hop loopers spend the winter in the adult stage, leaving hop yards in autumn to seek shelter elsewhere. The adult moths have been found overwintering (in low numbers) in caves and probably also exploit other protected sites such as cracks and crevices in tree trunks and fallen logs. They are thought to be capable of dispersing a considerable distance (several kilometers) to and from their overwintering sites, but the maximum extent of their migratory flight is unknown. The adults move back to hop yards in spring (early April), and eggs may be found on the hop foliage from mid-April onward. The eggs are initially slightly flattened and are roughly circular, with a mean diameter of 0.5–0.6 mm. Eggs are translucent and when first deposited have a faint greenish tinge that gradually disappears as the eggs mature. In the majority of cases, only a single egg is found on each leaf, although two or three are not uncommon and as many as 11 have been recorded. The eggs are always laid singly, not in groups, and are usually found on the underside of the leaf, either alongside a major vein (often at the point where the leaf blade joins the petiole) or on the leaf margin. Each adult female can lay up to 600 eggs during its lifespan (3–4 weeks for the summer generations). At 26°C, the eggs hatch in approximately 3 days. There are normally five larval instars, although some individuals undergo an additional (sixth) instar. At 26°C, it takes 14–15 days from egg hatch to pupation and an additional 9 days from pupation to adult eclosion. In the field, some larvae pupate on the leaves, while others pupate in the surface litter or just beneath the soil surface.

Data from adult trap catches and egg and larval samples suggest that there are three generations per year. However, the spring flight (and subsequent egg deposition) extends over a period of several weeks, so that from early May onward virtually all larval stages are present in the field at the same time, making it difficult to determine the optimal time at which to apply pesticides.

Damage

Large numbers of larvae can defoliate hop plants. Attacked leaves take on a characteristic "lacey" appearance. Such damage often appears to be particularly severe toward the base of the plant, although eggs are deposited with equal frequency at all heights in the canopy. Of more concern, however, are the losses that can result from direct damage to the cones (Fig. 112) at the end of the season when larval numbers tend to be highest.

Natural Enemies

Nine species of parasitoid have now been found associated with different looper life stages. These include two species of *Trichogramma* wasps (*T. deion* Pinto & Oatman and *T. thalense* Pinto & Oatman) that attack the egg stage. As many as three adult wasps can emerge from each hop looper egg, and season-long parasitism levels average about 20%, although peaks of up to 70% parasitism can occur at times.

At least five tachinid species parasitize the larval stages. As with the egg parasites, the level of parasitism is generally higher in the second and third generations than in the first, although even then only approximately 30% of the larvae are parasitized. A European species, *Compsilura concinnata* (Meigen), originally introduced for the control of gypsy moth (*Lymantria dispar* L.), appears to be the most abundant species. *C. concin-*

nata typically attacks the later instars (mainly fourth and fifth and occasionally third instars).

Hop looper pupae are attacked by two Ichneumonid wasps, *Pimpla sanguinipes* Cresson and *Vulgichneumon brevicinctor* (Say), but at present little information is available on their abundance or seasonal impact.

In addition to these parasitic species, generalist predators (e.g., vespid wasps and spiders) have been shown to be capable of exerting considerable mortality on looper larvae. Birds do not appear to have a significant effect on larval numbers, although their effect on adult populations is unknown.

Other Lepidoptera Found on Hop

Of the other Lepidoptera found on hop in the United States, only Bertha armyworm (*Mamestra configurata* Walker) appears to be common, and it can occasionally reach damaging levels. It differs from hop looper in having a much broader host range, and it lays its eggs in groups rather than singly. The first instar larvae can be distinguished from those of the looper by their black heads (which become paler in later instars) and by their aggregated distribution. Later instars have slightly more cylindrical bodies than do loopers and an additional pair of prolegs. Other Lepidoptera collected and reared from hop in Washington State are listed (Table 8), along with a subjective assessment of their relative abundance.

Management

No economic threshold has been established for hop loopers. Determining the optimal time to apply insecticides is difficult since for most of the season the larval population consists of a mixture of virtually all instars and at no point are the second and third generations dominated by small larvae. However, trials suggest that even the larger instars are readily controlled by a range of insecticides, at least at present. In order to avoid disrupting biological control of spider mites and aphids, the use of broad-spectrum compounds such as certain pyrethroids should be avoided. *Bacillus thuringiensis* subsp. *aizawai* Bonnefoi has proved effective in field trials and has the advantage of being specific to Lepidoptera pests.

(Prepared by T. Grasswitz)

Fig. 112. Damage caused by hop looper to cones and shoots. (Courtesy D. G. James)

Table 8. Leptidopteran larvae found on hop in south-central Washington (2005–2007)

Species	Relative abundance on hop
Mamestra configurata Walker (Bertha armyworm)	Common
Anavitrinella pampinaria (Guenée) (Common gray)	Fairly common (heavily parasitized)
Spilosoma sp.	Fairly common (heavily parasitized)
Lacanobia subjuncta (Grote & Robinson) (Speckled cutworm/Lacanobia fruitworm)	Occasional
Strymon melinus (Gray hairstreak)	Occasional
Trichoplusia ni (Hübner) (Cabbage looper)	Occasional
Xestia c-nigrum (L.) (Spotted cutworm)	Occasional
Udea profundalis (Packard) (False celery leaftier)	Rare

Root Weevils

Numerous species of root weevil, *Otiorhynchus* spp. (Coleoptera: Curculionidae), infest hop. The black vine weevil, *O. sulcatus* F., is the dominant species infesting hop followed by the strawberry root weevil, *O. ovatus* L., rough strawberry root weevil, *O. rugosostriatus* (Goeze), and *O. meridionalis* (Gyllenhall).

Biology and Life History

While the habits and characteristics of each weevil species are unique, the general life history for all species is similar. Pupation begins in mid-April, and adult emergence is generally complete by mid to late May. Oviposition by newly eclosed adults begins in late June and continues through September. Overwintered adults can begin to oviposit by late May and continue until early July. Adults are parthenogenic (only females are known) and therefore do not need to mate in order to lay eggs. Adults are nocturnal and begin to feed within 24 h of emergence from the soil. The preoviposition period (period from adult emergence to the onset of egg laying) averages 26 days on hop. Each adult can lay an average of 300 eggs (this number ranges from 12 to 1,230) in a single growing season. Dispersal of adults takes place via walking or transport by human activity. Black vine weevil adult movement in hop is fairly limited. Most adults remain in the hop hill that they fed on as larvae, particularly early in the growing season. As the season progresses, root weevils have been shown to travel distances of up to 35 m from their site of emergence. Root weevils overwinter in the soil, primarily as last instars, although a small percentage (5–10%) of the population can survive the winter in the soil as adults (Figs. 113 and 114).

The adult black vine weevil is typical of other weevils in that the head is elongated into a snout. They measure 8–10 mm in length and are jet black with golden-metallic patches of

Fig. 113. Fifth instar black vine weevil and feeding damage on rhizomes. (Courtesy D. Bruck)

elongated hair. The elytra are fused, preventing the adults from flying. Larvae are white, legless grubs with tan to dark brown head capsules.

Damage

Larvae are the most damaging stage; they feed from late summer into the following spring. Severe larval infestations can significantly shorten the life of a hop yard by feeding on rhizomes, roots, and crowns of plants. Larval feeding reduces plant vigor and can create wounds that serve as entrance points for soilborne plant pathogens. The most severe damage occurs in the spring when late instars are observed scoring and girdling hop roots that may range from 3 to 20 mm in diameter. Root damage by larval feeding results in water stress and reduced nutrient uptake and plant growth. Yards heavily infested with root weevil larvae can drop leaves prematurely and may require that individual plants or entire yards be removed from production. Leaf notching is also a diagnostic symptom of adult activity and signals the onset of preovipositional feeding in the spring. Sampling for leaf notching can be useful in helping time insecticide applications.

Management

The adult stage of black vine and other root weevils can be controlled with insecticide applications. It is essential that adult weevils be targeted while they still are in their preovipositional period. Eliminating these adults before they begin to lay eggs prevents reinfestation. Insecticide applications are effective only when applied at night when adult weevils are active. Adult weevil populations can be monitored by using grooved boards and pitfall traps to determine when adults are active in the spring or by examining plants for the characteristic leaf notching. Applying an insecticide 2–3 weeks after the emergence of the first adult allows for more adults to be targeted with the first spray.

Black vine weevil larvae have been controlled successfully in hop with the use of both Heterorhabditid and Steinernematid nematodes. Application of nematodes significantly reduced (by 70%) the number of living adult black vine weevils in hop yards. In order to be effective, nematode applications must be timed properly to coincide with the presence of late instars, soil temperatures above 12°C, and adequate soil moisture. Several species of root weevil larvae can also be controlled with the use of the entomopathogenic fungi *Metarhizium anisopliae* (Metschn.) Sorokin and *Beauveria bassiana* (Bals.-Criv.) Vuill. *M. anisopliae* is very effective at controlling larval infestations in small fruit crops and ornamental nursery plants.

Selected References

Baird, A. R., Baird, C. R., and Dorschner, K. W. 1992. Dispersal of black vine weevil adults, *Otiorhynchus sulcatus* (F) in Idaho hop yards (Coleoptera: Curculionidae). J. Idaho Acad. Sci. 28:8-10.

Baird, C. R., Dorschner, K. W., and Nyberg, C. J. 1992. Biology of the black vine weevil *Otiorhynchus sulcatus* on hop in Idaho (Coleoptera: Curculionidae). J. Entomol. Soc. B. C. 89:31-37.

Bruck, D. J. 2005. Ecology of *Metarhizium anisopliae* in soilless potting media and the rhizosphere: Implications for pest management. Biol. Control 32:155-163.

Dorschner, K. W., Agudelo-Silva, F., and Baird, C. R. 1989. Use of Heterorhabditid and Steinernematid nematodes to control the black vine weevils in hop. Fla. Entomol. 72:554-556.

Moorhouse, E. R., Charnely, A. K., and Gillespie, A. T. 1992. Review of the biology and control of the vine weevil, *Otiorhynchus sulcatus* (Coleoptera: Curculionidae). Ann. Appl. Biol. 121:431-454.

Smith, F. 1932. Biology and control of black vine weevil. U.S. Dep. Agric. Tech. Bull. 325:1-43.

Fig. 114. Black vine weevil life stages (left to right): egg; first, second, third, fourth, fifth, and sixth instars; and pupa. Bar = 1 mm. (Courtesy D. Bruck)

(Prepared by D. Bruck)

Rosy Rustic Moth

Rosy rustic moth, *Hydraecia micacea* (Esper) (Lepidoptera: Noctuidae), is an occasional pest of hop and other cultivated crops in Europe. Its primary habitat is wetlands. The polyphagous stem-boring larvae attack a wide range of host plants, including beet, corn, cereals, hop, onion, potato, strawberry, and tomato. The caterpillars are frequently referred to by their common name of "potato stem borer." The species also infests a number of weeds, especially quackgrass, *Elytrigia* (syn. *Elymus*, *Agropyron*) *repens* (L.) Desv. ex Nevski, docks and sorrels (*Rumex* spp.), and lambsquarter (*Chenopodium album* L.).

A native of the northern Palaearctic region, the rosy rustic moth has a distribution from the arctic circle southward to Spain and eastward to Japan. In the early 1900s the rosy rustic moth was introduced to the maritime provinces of eastern Canada. It was first identified in the United States from specimens collected in Massachusetts in 1917. More recently it has been confirmed as a pest of corn in the Midwest. To date it has not been collected west of the Rocky Mountains and has not yet been confirmed as a pest of hop in the western United States.

Biology and Life History

Moths of *H. micacea* have a wide variation in color and size and possess a wingspan of about 30–45 mm. The forewings vary from light to dark rosy brown. The hind wings have a glossy appearance and gray veins and range in color from pale white to light fuscous.

First instar larvae are normally dark purple. Subsequent instars turn progressively from a pale purple to pink, and last instars are grayish white. The ventral side is always whitish, and a dark band crosses each abdominal segment dorsally in all instars. Each segment carries a similar number of dark tubercles that each bear one or two bristles. The head capsule is glossy red-brown, and the pronotum is yellow to light brown with a dark margin. Full-grown larvae may reach a length of 50 mm.

The rosy rustic moth has one generation per year. In late summer, moths emerge from pupae in the soil and mate within a few days. Peak emergence occurs during August. The moths are crepuscular and inactive during the day. Adults fly and disperse from late summer through October. The mean number of eggs laid per female is 100–150. Eggs are deposited in small groups on or about the crowns of host plants or scattered nearby on the soil surface. Quackgrass seems to be a preferred host. The eggs can tolerate very low temperatures and require a minimum of 2 months of near freezing temperatures in order to terminate diapause and induce high percentages to hatch. The eggs hatch from approximately mid-April to mid-May. The larval stage lasts up to 10 weeks. Final instar larvae leave the host plant, burrow to a depth of 5–10 cm, and construct cells in which to pupate. Within 3–4 weeks, moths of the next generation begin to emerge from mid-July onwards.

Damage

The rosy rustic moth is not a common hop pest in Europe and only occasionally causes significant damage. The larvae feed underground on roots and crown tissue (Fig. 115) and also act as stem borers within the bines and stalks of host plants. Feeding injury on hop results in bine dieback and reduced cone production. Bines injured by stem-boring *H. micacea* larvae also have a greater incidence of Fusarium canker than undamaged bines.

Attacks on hop are most prevalent close to water or in moist areas with high water tables and on light, sandy soils. Serious damage is usually restricted to the headlands and borders of hop yards and usually those yards with significant numbers of uncontrolled weed hosts. Young larvae of *H. micacea* feed on these weed hosts for a period of time in the spring and then may leave weed hosts as second instar or larger larvae in favor of hop plants. Larvae make small entrance holes in the shoots near the soil level. In June, after about a month of feeding as stem borers, the third and later instars descend to tunnel or feed externally on crowns and roots. First instar larvae that hatch from eggs directly deposited on or near hop crowns and bore into hop shoots seldom establish economic infestations, because they are removed or killed when crowns are pruned in early spring. Larvae that subsequently migrate from weeds to hop bines cause the most injury.

Damage to hop plants first becomes visible in April and May, when young shoots can wilt and desiccate. Young hop plants are especially vulnerable. Early instar larvae tunneling the pith of bines are usually less easily detected, since the infested plants rarely show symptoms of wilt because the plant's vascular system is not actually damaged. As spring advances, larvae tend to disperse from their initial host plant to neighboring hop plants, thereby infesting sometimes three or four plants prior to reaching maturity. Larval exit holes are visible on bines near the soil level and sometimes cause bines to split, especially when the hole is in contact with the training wire. However, probably the most severe impact of rosy rustic moth to hop occurs from larvae feeding below ground during summer. Up to 30 larvae per plant may be found feeding on the roots, which can lead to the complete destruction of the root system and is sometimes noticed only in the following year.

In North America this species has been of primary economic significance as a pest of corn. However, cultivated hop may be at risk should the pest be introduced into hop-producing regions of the western United States.

Management

In consideration of the biology of rosy rustic moth outlined above, the priority control measure to limit infestation is to maintain hop yards free of weeds, especially of those species mentioned above. Severe damage is always associated with weedy hop yards. Because the larvae are stem borers as well as external feeders of crown and roots below the soil surface, chemical control is difficult. A registered systemic insecticide watered onto the bases of plants, preferably during May or June, may help control the stem-boring stages.

Fig. 115. Larvae of the rosy rustic moth on a shoot. (Courtesy J. Vostrel)

Selected References

French, N., Ludlam, F. A. B., and Wardlow, L. R. 1973. Biology, damage and control of rosy rustic moth, *Hydraecia micacea* (Esp.), on hops. Plant Pathol. 22:58-64.

Giebink, B. L., Scriber, J. M., and Hogg, D. B. 1985. Developmental rates of the hop vine borer and potato stem borer (Lepidoptera: Noctuidae): Implications for insecticidal control. J. Econ. Entomol. 78:311-315.

Scherney, F. 1970. *Hydroecia micacea* Esp. als Schädling an Hopfen und Mais. Ges. Pflanz. 22:106-108.

Šedivý, J., Born, P., and Vostřel, J. 2005. Harmful occurrence of rosy rustic moth (*Hydraecia micacea*) (Noctuidae, Lepidoptera) on hop in the Czech Republic. Plant Prot. Sci. 41:150-157.

Theobald, F. V. 1928. The rosy rustic moth (*Hydroecia micacea*, Fab.) as a hop pest. J. S. E. Agric. Coll. Wye 25:71-74.

Tölg, F. 1911. *Hydroecia micacea* Esp., ein neuer Hopfenschädling. Landeskulturrat für das Königreich Böhmen, Saaz [Žatec].

West, R. J., and Laing, J. E. 1985. Development of the potato stem borer, *Hydraecia micacea* (Lepidoptera: Noctuidae) in the laboratory and field. Proc. Entomol. Soc. Ont. 115:81-87.

(Prepared by F. Weihrauch and G. C. Fisher)

Two-Spotted Spider Mite

The two-spotted spider mite is a serious pest of hop and numerous other agronomic and ornamental plants. It is distributed worldwide and reported to feed on more than 180 plant species in both field and greenhouse environments.

Biology and Life History

Spider mites (*Tetranychus urticae* Koch) are related to spiders and ticks. Adult spider mites are very small and oval with four pairs of legs. Females are 0.4–0.6 mm long and yellow to yellow green and have a large dark spot on each side of the abdomen. These dark spots, present on both sexes, represent food ingested by the mites that is visible through the body, and they become more pronounced as mites mature. The male is about three-fourths the size of the female, yellower, and has a more pointed abdomen posteriorly. Two-spotted spider mites overwinter as diapausing (seasonally dormant) adult females on dead vegetation, soil debris, hop crowns, cracks and crevices on hop poles, and other protected areas in and adjacent to hop yards. Male two-spotted spider mites do not overwinter. Females emerge from overwintering sites as the weather warms in early spring, move to young foliage, and immediately begin feeding.

There are five stages of development: egg, larva, protonymph, deutonymph, and adult. Female mites can lay up to 16 eggs per day and 50–240 during their lifetime. Spider mite eggs are clear to pearly white spheres about 0.13 mm in diameter and are usually laid on the undersides of leaves. Eggs can give rise to either male or female spider mites. Female spider mites emerge from fertilized (diploid) eggs, while males emerge from unfertilized (haploid) eggs. The first-stage immature mites (larvae) are about the same size as eggs and have only three pairs of legs. Larvae are initially translucent but become yellow green to green as they begin feeding. The dark spots on the abdomen may not be particularly pronounced. Male and female larvae cannot be distinguished accurately. Protonymphs and deutonymphs are progressively larger than the larvae. In these developmental stages, two-spotted spider mites are more deeply green and the characteristic black spots on the abdomen are more pronounced as compared to immature stages (larvae). Male deutonymphs can be distinguished from females by their smaller size and more pointed abdomen (Fig. 116).

The time needed for development of spider mites from egg to adult depends on temperature. No development occurs below about 10°C or above about 38°C. Between these lower and upper developmental thresholds, the time needed to develop from egg to adult decreases with temperature such that about 36 days are required at 13°C and only 6 days are required at 28°C. Five to eight overlapping mite generations develop on hop during the spring and summer months. In the summer, development from egg to adult may take only 7–10 days. Spider mites thrive under hot, dry conditions, and large colonies of mites produce substantial webbing of leaves, which aids in protection from predators. Mites can also spread by wind transport, particularly when populations are high. In late August and September, when the day length shortens, mites become orange without the dark spots and begin moving down hop bines to overwintering sites.

Damage

Two-spotted spider mites have a wide host range, feeding and reproducing on more than 180 plant species, and they are important pests of many field, forage, ornamental, and horticultural crops. Two-spotted spider mites feed on plants by piercing the leaf tissue with their mouthparts and withdrawing plant juices from the cells. Spider mites damage hop plants by feeding directly on leaves and cones. Leaf feeding by spider mites results in silvering and bronzing of leaves. In heavy infestations the spots coalesce and leaves become yellow to brown and desiccate, which reduces plant vigor. Severe infestation can cause complete defoliation and is accompanied by heavy production of webbing that can hinder chemical and biological control efforts (Fig. 117). Most economic damage, however, is associated with cones. Spider mites not only contaminate the cones by their presence, but feeding on cones results in desiccated, brittle, and discolored (red) cones (Fig. 118). Cones damaged by spider mites tend to shatter so that both quality and quantity of yield is reduced. Cones damaged by spider mites may have accelerated oxidation after harvest and reduced storagability. When infestations are severe, total crop loss can occur.

Management

An integrated approach to spider mite management should include monitoring for spider mite presence and population density (i.e., the number of spider mite motile stages per leaf), the use of economic thresholds to indicate the existence of economically damaging spider mite densities, and effective treatment options (biological, chemical, and cultural) for maintaining populations below the economic threshold. Low to moderate numbers of mites on hop foliage may be tolerated if the weather is mild and sufficient biological control agents are present. However, spider mite populations can build very rapidly, especially in hot, dry conditions, and it is important to monitor the presence of spider mites and their natural enemies

Fig. 116. Male (left), nymph with egg (center left), and females (right two images) two-spotted spider mites. (Courtesy D. H. Gent, T. Neill, J. D. Barbour, and W. F. Mahaffee, left to right)

Fig. 117. Chlorosis, death, and webbing of portions of leaves associated with spider mite feeding damage. (Courtesy D. H. Gent, left, and D. G. James, center and right)

to determine the need for treatment. Hop foliage should be monitored at least weekly by checking the undersides of leaves for spider mites, mite eggs, and webbing, as well as stippling and yellowing of leaves associated with spider mite feeding. The average number of motile spider mites per leaf should be determined from a representative sample of leaves collected from the field. A 10× hand lens is useful for this purpose.

Hop yards should be treated only if pest populations reach or exceed established economic thresholds to minimize input costs, delay development of resistance, and conserve natural enemy populations. Unfortunately, a comprehensive economic threshold based on spider mite economic injury levels has not been developed for spider mites on hop. Many growers use a nominal or provisional threshold of five to 10 spider mites per leaf; however, research in Germany indicates that hop plants may be able to tolerate much higher spider mite populations without suffering economic loss. When a miticide is chosen, its safety to biological control agents as well as its efficacy should be considered. A number of miticides available for spider mite control on hop are reported to be relatively safe to predatory insects and mites. The use of these less toxic materials enhances biological control and prevents overuse of chemical control.

Spider mites have an extensive community of natural enemies, for example, predatory mites, big-eyed bugs, minute pirate bugs, lady beetles (mite destroyers), spiders, and lacewings, that may regulate populations successfully when undisturbed by pesticides or cultural practices. Spider mite populations can be exacerbated by the use of certain pesticides (e.g., pyrethroid insecticides and multiple applications of sulfur), which can enhance spider mite reproduction or suppress spider mite predators. Spider mite outbreaks may result from foliar applications of organophosphate, carbamate, pyrethroid, and neonicotinoid compounds used to control spider mites and other pests such as aphids and caterpillars. Use of these pesticides should be avoided where possible. Preservation of endemic spider mite natural enemies in hop yards allows biological control to play a large role in management, reducing the need for chemical control. Adoption of conservation biological control in hop yards in Washington State appears to have reduced miticide use from two to three applications to one application per season. Recruitment of predators to hop yards using herbivore-induced plant volatile attractants (e.g., methyl salicylate) is showing promise in improving biocontrol in hop yards.

Cultural control practices usually modify the relationships between a pest population and its natural environment and are sometimes known as ecological control methods. In the case of spider mites, hot, dry conditions alone or in combination with stressed plants, particularly water-stressed or dust-covered plants, favor mite outbreaks. Spider mites develop and reproduce more quickly and feed more under dry conditions. Many of the spider mite's natural enemies are stressed and provide less control under these conditions. In addition, drought-stressed plants are often nutritionally superior to adequately watered plants for mite development and reproduction. Reducing plant stress by providing adequate fertilizer and irrigation is therefore a key factor in managing spider mite populations. Spider mite problems are often more severe on dust-covered plants, and many practitioners feel that dust control is an important mite-prevention practice. It is not clear whether dust increases spider mite populations directly by drying the leaf surface or enhancing web attachment or indirectly through changes in plant chemistry caused by stress or by reducing natural enemy effectiveness. Dust can be suppressed by covering dirt roads with gravel, straw, or crop debris, by watering or oiling roads, and by reducing driving speed.

Planting ground covers can also reduce mite problems by providing a microclimate more favorable to the spider mite's natural enemies and reducing conditions leading to dust-stressed plants. The use of ground covers can minimize dust and provide a refuge with a microclimate favorable for the spider mite's natural enemies.

Selected References

Gardiner, M. M., and Barbour, J. D. 2003. Arthropod diversity and abundance on feral and cultivated hop, *Humulus lupulus* (Urticales: Cannabaceae), in Idaho. Environ. Entomol. 32:564-574.

James, D. G. 2002. Selectivity of the miticide, bifenazate and aphicide, pymetrozine, to spider mite predators in Washington hops. Int. J. Acarol. 28:175-179.

James, D. G. 2003. Pesticide susceptibility of two coccinellids (*Stethorus punctum picipes* (Casey) and *Harmonia axyridis* Pallas) important in biological control of mites and aphids in Washington hops. Biocontrol Sci. Technol. 13:253-259.

James, D. G., and Price, T. S. 2002. Fecundity in twospotted spider mite (Acari: Tetranychidae) is increased by direct and systemic exposure to imidacloprid. J. Econ. Entomol. 95:729-732.

James, D. G., and Price, T. S. 2004. Field-testing of methyl salicylate for recruitment and retention of beneficial insects in grapes and hops. J. Chem. Ecol. 30:1613-1628.

Strong, W. B., and Croft, B. A. 1995. Inoculative release of phytoseiid mites (Acarina: Phytoseiidae) into the rapidly expanding canopy of hops for control of *Tetranychus urticae* (Acarina: Tetranychidae). Environ. Entomol. 25:446-453.

Weihrauch, F. 2005. Evaluation of a damage threshold for two-spotted spider mites, *Tetranychus urticae* Koch (Acari: Tetranychidae), in hop culture. Ann. Appl. Biol. 146:501-509.

Wright, L. C., and Cone, W. W. 1999. Binomial sequential sampling plans for adult female twospotted spider mites (Acar:Tetranychidae) on hops. J. Econ. Entomol. 92:1335-1343.

(Prepared by D. G. James and J. D. Barbour)

Fig. 118. Red discoloration of cones characteristic of spider mite feeding damage. (Courtesy D. H. Gent, left, and D. G. James, center and right)

Insect Pests of Minor Importance

White Grubs

Biology and Life History

White grubs are the larvae or immature stages of scarab beetles (Scarabaeidae), two or more species of which may feed on and injure the roots in commercial hop yards. Adults are commonly called May and June beetles because they are active during the spring and summer. They are attracted to light and are often seen flying around well-lit homes on warm summer evenings. Species of *Polyphylla* (the ten-lined June beetle, *P. decemlineata* Say, and the coastline June beetle, *P. crinita* LeConte) are large (25–35 mm), tan beetles with white longitudinal stripes on the pronotum and elytra (wing covers) (Fig. 119). Adult *Phyllophaga* spp. are metallic reddish-brown beetles about 2 cm long. Larvae of all species are soil-dwelling, cream-colored, C-shaped grubs with brown heads and strong mandibles. The larvae range in size from just 2–3 mm at hatching to 50 mm at maturity and have three pairs of conspicuous legs, used for moving through the soil, attached to the thorax (Fig. 120).

May and June beetles have a 3-year life cycle. Beetles mate within a few days of emerging from the soil during May and June. Females lay eggs in the soil that hatch in 3–4 weeks. The emerging grubs feed on plant roots near the soil surface during the summer, moving below the frost line to avoid freezing conditions as winter approaches. The following spring, larvae return to the root zone and continue to feed throughout the summer. As winter approaches, the grubs complete development, cease feeding, pupate, and then become inactive adults within the pupal case. Adults emerge during the spring and summer of the following year.

Damage

Adult beetles feed on the leaves of conifers and broadleaf trees and shrubs but seldom cause economic damage. The larvae feed on plant roots and are major pests of turfgrasses and grain crops but also attack the roots of small fruits, tree fruits, ornamentals, and field crops including hop. Feeding on roots by larvae over multiple seasons destroys root reserves, damages conductive tissue, and causes bines to wilt and die, symptoms similar to those caused by California prionus. Plants growing in sandy soils appear to be especially susceptible, and the most damage is caused by the large, second-year larvae.

Management

No insecticides are registered for control of white grubs in hop, but white grub infestations generally do not warrant treatment. Infestations are generally localized within yards and spread slowly from the site of initial infestation. Infestations are difficult to detect and are often first noticed when small groups of plants, usually restricted to just a few hills, wilt and die. Infestations can be confirmed by digging up and inspecting roots for damage and the presence of grubs. Control requires the removal of infested plants.

Skunks, raccoons, and rodents feed on white grubs; however, the damage caused by these mammals in search of food may exceed that caused by white grubs. The grubs are also attacked by parasitic nematodes in the genus *Heterorhabditis*. Commercial preparations of *Heterorhabditis* spp. have been used with limited success to control white grubs in a variety of crops.

Plants showing symptoms of infestation should be removed from hop yards, and roots and soil should be inspected for the presence of larvae. If infestation is confirmed, removing the infested plant and one or two plants on either side may help to slow the spread. Yards established on or near favored larval hosts are at an increased risk. New yards on or near former sod areas and grain crops should be avoided. Soil tillage prior to planting may help reduce the population by killing the larvae or exposing larvae to natural predators and parasites. When practical, planting a nonhost cover crop or rotational crop, such as clover, may help reduce the population. Avoiding conditions that lead to drought stress may reduce injury in infested yards.

Selected References

Berry, R. E. 1998. Insects and Mites of Economic Importance in the Northwest, 2nd ed. OSU Bookstore, Corvallis, OR.

Van Steenwyk, R. A., and Rough, D. 1989. Biology of the tenlined June beetle (Coleoptera: Scarabaeidae). J. Econ. Entomol. 82: 1738-1742.

(Prepared by J. D. Barbour)

Fig. 119. Adult June beetle. (Ken Gray Image courtesy Oregon State University)

Fig. 120. Larva of the June beetle. (Ken Gray Image courtesy Oregon State University)

Western Spotted Cucumber Beetle

Biology and Life History

Adult *Diabrotica undecimpunctata* Mannerheim are small (6–9 mm), yellowish-green beetles with 11 distinct black spots on the wing covers (elytra) (Fig. 121). Fertilized females overwinter and may be active on warm winter days. Eggs are deposited near the bases of host plants in early spring and hatch in 7–10 days, depending on temperature. A single female can lay between 200 and 1,200 eggs. Larvae immediately begin feeding on plant roots and grow from less than 2 mm long to 14–17 mm long in 2–4 weeks. A mature larva is white with a brown head and a sclerotized plate on the dorsal side of the terminal abdominal segment (Fig. 122). Larvae complete development in the soil and pupate in late spring, and new adults are produced by early July in western Oregon. The complete life cycle requires 30–60 days. There are two generations per year in the Pacific Northwest of the United States.

Damage

The adult of the western spotted cucumber beetle is the stage that may damage hop plants. Beetles prefer tender, succulent portions of plants and feed on leaves, flowers, and fruits of many plant species including hop. Adults may also attack seedlings and young shoots, girdling stems and killing young plants. Adult feeding is not generally of economic importance in hop, except when growing tips of young plants or developing hop flowers are attacked. Root feeding by larvae has not been reported as an economic problem on hop.

Management

Hop is not a favored host of the western cucumber beetle, and it is seldom attacked in numbers warranting management efforts. No insecticides are registered for control of the western spotted cucumber beetle on hop.

Crop rotations and delayed planting strategies used to manage cucumber beetles in annual field and horticultural crops are not applicable to hop production. Preventing establishment of weed hosts in yards and yard borders may reduce the risk of attack. Hop yards near favored larval hosts such as cucurbits and corn may have a higher risk of attack by adult beetles.

Western cucumber beetles are attacked by a variety of natural enemies, including generalist predators and parasitoids (e.g., the tachinid fly, *Celatoria diabroticae* [Shimer]). Avoiding unnecessary use of broad-spectrum pesticides may help to preserve natural enemies.

Fig. 121. Adult cucumber beetle. (Courtesy D. H. Gent)

Fig. 122. Larva of the cucumber beetle. (Ken Gray Image courtesy Oregon State University)

Selected Reference

Berry, R. E. 1998. Insects and Mites of Economic Importance in the Northwest, 2nd ed. OSU Bookstore, Corvallis, OR.

(Prepared by J. D. Barbour)

Wireworms

Biology and Life History

Wireworms are the larvae of click beetles (Elateridae; *Agriotes* spp., *Limonius* spp., and *Ctenicera* spp.). Adults are relatively slender, brown to gray beetles (Fig. 123) that are common in the summer and are often attracted to lights on warm nights. Females emerge in the spring from wintering sites and lay eggs in the soil near the bases of host plants. Each female can lay up to 300 eggs, which hatch in 1–4 weeks, depending on soil temperature and the wireworm species. Early instar larvae feed on the roots of plants and soilborne fungi. Later instars are hard-bodied, slender, cylindrical, yellow-brown "worms" that feed on roots, rhizomes, and tubers of plants. Wireworms may reach 3 cm in length, and, although they have flexible bodies, they are relatively tough and resistant to crushing. They have one pair of short legs on each of the three segments immediately behind the head. The terminal segment is either pointed or has paired hooks separated by a central gap (Fig. 124). Larvae live

Fig. 123. Adult wireworm. (Ken Gray Image courtesy Oregon State University)

Fig. 124. Larva of the wireworm. (Ken Gray Image courtesy Oregon State University)

in the soil for 3–5 years, moving down in the soil profile to avoid freezing during the winter and excessive heat (>27°C) or desiccation during the summer. Most feeding occurs within 15 cm of the soil surface. Life cycles vary, but typically, mature larvae pupate during the late summer and fall, becoming adults that overwinter deep in the soil and emerge the following spring. Eggs and young larvae are preyed upon by ground beetles, and larvae often succumb to infections by naturally occurring fungi.

Damage

Adults cause no damage. Wireworm larvae attack roots, bulbs, and tubers of many crops, but they are particularly destructive to cereal crops and potato. Although few crops escape damage, economic losses resulting from wireworm feeding on the roots of hop plants is uncommon.

Management

As is the case for most subterranean pests, existing wireworm infestations are difficult to manage. Planting in uninfested soil is the best management practice. Wireworm larvae are attracted to odors emitted by fermenting grain seeds. The presence of larvae in the soil can therefore be monitored by counting the number of larvae collected at buried bait stations containing germinating wheat, barley, or oat seeds. Baiting must be carried out when soil temperatures are 10–15°C to ensure that wireworms are active near the surface of the soil, and at least 10 stations should be used in each yard. No economic thresholds are available for hop; however, when the average number of wireworms per baited trap exceeds one, damage to young plants may occur. No insecticides are registered for control of wireworms on hop.

New hop yards should not be planted in a field that has a history of previous wireworm infestation or a field previously planted to wireworm-susceptible hosts such as grains, sod or grassy pastures, or pastures with grass weeds, without first determining the potential for wireworm problems (e.g., by monitoring with buried bait stations). In fields known to contain wireworms, tillage may reduce numbers. Fields with heavy infestations should be fallowed or planted to nonhost crops (e.g., alfalfa) for several years. Planting or inter-planting rape or mustard crops containing high levels of glucosinolates (plant compounds whose breakdown products are toxic to many insects) and plowing in the green crop may reduce wireworm numbers or wireworm feeding.

Selected References

Berry, R. E. 1998. Insects and Mites of Economic Importance in the Northwest, 2nd ed. OSU Bookstore, Corvallis, OR.

Parker, W. E., and Howard, J. J. 2001. The biology and management of wireworms (*Agriotes* spp.) in potato with particular reference to the United Kingdom. Agric. For. Entomol. 3:85-98.

Ratcliffe, S. T., Gray, M. E., and Steffey, K. L. 2004. Wireworms, Family Elateridae. http://www.ipm.uiuc.edu/fieldcrops/insects/wireworm/

(Prepared by J. D. Barbour)

Part III. Postharvest Disorders and Diseases

Bale Self-Heating

Self-heating, or reheating, and the resulting spontaneous combustion of bales is a common, expensive, and difficult problem for the hop industry (Fig. 125). Recent examples of bale combustion include warehouse fires in Washington in 1993, 1999, 2000, and 2006 and in Oregon in 1983. The fire in 2006 destroyed approximately 4% of the United States' supply at that time. Little was known about the cause and management of self-heating at the time of the 1999 and 2000 warehouse fires. Bale self-heating is now thought to be associated with inadequate or excessive drying of hops. In the presence of sufficient moisture and oxygen, oxidation of resins and organic matter results in an exothermic and self-sustaining reaction that can cause ignition of cones if temperatures become sufficiently high. Bale self-heating can occur with any cultivar, but it is particularly problematic with "super-alpha" cultivars, such as Columbus, Tomahawk, and Zeus, which are prone to rapid oxidation and instability in storage.

The risk of bale self-heating can be minimized by careful drying, cooling, and baling. Cones should be dried to 8–10% moisture content. Drying at temperatures greater than 66°C increases the risk of bale self-heating, reduces crop quality and alpha-acid content, and is unnecessary. Following correct cooling and conditioning procedures is important. Large volumes of ambient air should be provided after the kiln heat has been turned off to fully equilibrate cone moisture content. Wrapping bales with low air-permeable polypropylene rather than burlap also reduces reheating. Bale temperature at receipt at a warehouse facility should not be greater than 29°C. Following these measures also will improve and preserve the quality and appearance of cones.

Selected References

Fric, V., Loukota, M., and Makovec, K. 1974. Vliv extremnich zpusobu osetrovani a suseni chmele na kvalitu chmele. Rostl. Vyroba 20:757-764.

Jones, J. C., and Raj, S. C. 1988. Self heating and ignition of hops. J. Inst. Brew. 94:139-141.

Fig. 125. Brown discoloration of cones within a bale caused by self-heating. (Courtesy D. R. Smith)

(Prepared by D. Hysert)

Cone Early Maturity

Cone early maturity is caused by infections of the strig, bracts, or bracteoles of cones by *Podosphaera macularis* (Wallr.) U. Braun & S. Takam. (see Powdery Mildew) after cone elongation has ceased. Disease development is usually associated with low to medium disease levels in a yard and with an unfavorable environment for disease to develop during burr and cone elongation, followed by conditions favorable for infection prior to harvest. Cone early maturity disease development is also common in yards with low disease levels but that are near those with medium to high levels of disease. The most common symptoms prior to harvest are early or rapid maturation (drying) of cones, which results in subtle loss of green color in individual bracts and bracteoles or the entire cone, and red-brown flecking of individual bracts and bracteoles (Fig. 126). Signs of these infections are generally not visible to the unaided eye and require dissection of cones for microscopic examination for hyphae, conidiophores, and conidia (Fig. 127). In regions that have both mating types of *P. macularis*, chasmothecia (small, red to brown spheres) are often visible. Symptoms may not be noticed until after kiln drying, when affected cones have red-brown discoloration (Fig. 127), altered aroma, and little or no green color. Abscission zones between

Fig. 126. Cones of cultivar Willamette exhibiting severe symptoms of hop cone early maturity. Note reduced green color and light brown discoloration of individual bracts and bracteoles. (Courtesy W. F. Mahaffee)

the rachis and bracteoles are associated with infected bracteoles or posterior to point of infection on the rachis (Fig. 128). The brown flecks visible on infected bracteoles are associated mainly with *P. macularis* hyphae; however, other fungi (*Botrytis* spp., *Alternaria* spp., and particularly *Fusarium* spp.) have also been observed in association with these flecks. Hop cultivars that have loose, open cones are more susceptible to developing cone early maturity than those with dense, compact cones. Losses from cone early maturity are associated with cone shatter during picking and cleaning, poor color and aroma after drying, and cone shatter and poor storage during and after baling. Cone early maturity is managed by controlling powdery mildew within the yard and being aware of disease pressure in surrounding hop yards (see Powdery Mildew). If late-season powdery mildew infections do develop, more frequent monitoring of changes in dry matter may indicate picking a few days earlier than normal, which can minimize loss of cone quality. However, an early harvest can also result in yield reduction.

Fig. 127. Dried and baled cones from Figure 126 with extensive medium brown discoloration. (Courtesy W. F. Mahaffee)

Selected References

Coley-Smith, D. J. 1964. An association between so-called premature ripening of hops and the presence of powdery mildew (*Sphaerotheca macularis*). Rep. Dep. Hop Res. Wye Coll. 1963:30-31.

Hammond, W. H. 1900. On red mould, or the premature 'going off' of hops. J. S. E. Agric. Coll., Wye 9:19-20.

(Prepared by W. F. Mahaffee)

Fig. 128. Microscopic signs and symptoms of early cone maturity. A, Infected cone; B and C, bracteole with reduced green color and brown flecking; D, strig with diffuse infections and abscission zones at the point of bract/bracteole attachment; E, bract with diffuse infection; and F, conidia and conidiophores intermixed with hairs along the strig. (Courtesy W. F. Mahaffee)

Noninfectious Abiotic Disorders

Nutrient Imbalances

Many of the symptoms associated with nutrient imbalances are not specific to any particular nutrient deficiency and may vary depending on environment and cultivar. They are also similar to the symptoms caused by several plant pathogens, particularly viruses. Therefore, leaf, petiole, or soil analyses are required for accurate diagnosis of nutrient disorders.

Boron

Symptoms of boron deficiency first appear as delayed emergence of shoots (Fig. 129). Once leaves do emerge, they have thick stems with short internodes. Leaves are small and distorted or crinkled and often lack lobes. Laterals develop early, and the growing tip often dies. Additional symptoms may include numerous buds on the crown at the ground, large stipules immediately behind the growing tip, and browning and death of burrs. Boron deficiency has been implicated as a contributing factor for red crown rot.

Calcium

When calcium is deficient, symptoms are first expressed in young tissues and later spread downward. Growing tips become yellow and die while young leaves are small and develop brown spots and marginal necrosis (Fig. 130). Excess calcium can result in deficiencies in other cations (e.g., magnesium, potassium, and iron) with associated symptoms. Abundance of calcium also can cause premature yellowing and senescence of cones.

Iron

Iron deficiency is first observed at the growing points of the bine with young leaves becoming chlorotic (Fig. 131). Leaves may have green tips with yellow bases.

Magnesium

Insufficient magnesium manifests as chlorosis of the interveinal areas of lower leaves followed by necrosis (Fig. 132). Occasionally, marginal scorching and early defoliation occur. Symptoms may later progress up the bine. Magnesium deficiency can also be caused by excessive potassium in the soil.

Molybdenum and Manganese

The symptoms of molybdenum and manganese deficiencies are interveinal necrotic spotting and leaf chlorosis. Availability of molybdenum and manganese becomes limited in soils with high pH (Fig. 133). Manganese accumulation in plant tissues increases at pH below 5.7 and may become toxic in some acidic soils.

Fig. 129. Small, distorted, and chlorotic leaves caused by boron deficiency. (Courtesy C. B. Skotland)

Fig. 130. Marginal chlorosis and necrosis caused by calcium deficiency. (Courtesy C. B. Skotland)

Fig. 131. Interveinal chlorosis of young leaves caused by iron deficiency. (Courtesy C. B. Skotland)

Fig. 132. Interveinal chlorosis of older leaves caused by magnesium deficiency. (Courtesy C. B. Skotland)

Fig. 133. Interveinal necrosis (left) and chlorosis (right) caused by molybdenum deficiency. (Courtesy C. B. Skotland; left and P. McGee; right)

Fig. 134. Light green appearance and necrosis of leaves caused by nitrogen deficiency. (Courtesy C. B. Skotland)

Nitrogen

When nitrogen is limited, the plant is usually stunted, has reduced flowering, and appears chlorotic. Symptoms first appear on lower leaves and later spread upward to younger leaves (Fig. 134). Leaves develop a pale yellow color, are thin, have narrow lobes and red petioles, and are shed early. The cones that do develop are small and underdeveloped.

Excess nitrogen results in abundant sucker growth, and cones become oversized and may develop "angel wings" (leaves developed from the cone rachis). Plants are also more susceptible to some diseases and insects because of the succulent growth.

Phosphorus

Phosphorus deficiency results in thin bines with downcurved, dark green leaves with a dull appearance and brown discoloration of the outer parts of cones. Symptoms first appear on lower leaves and spread upward. Excessive phosphorus fertilization may induce zinc deficiencies, particularly in soils that are marginally deficient in zinc.

Potassium

Hop has a high potassium requirement. Deficiency results in weak bine growth and reduced burr formation. Young leaves develop a bronze color in the interveinal areas, which become necrotic with an ash gray appearance (Fig. 135). Older leaves sometimes develop marginal scorching and drop prematurely. Symptoms are more marked when nitrogen levels are high.

Sulfur

Sulfur can be deficient in some soils, particularly acidic, coarse-textured soils, resulting in stunted growth, spindly stems, and chlorosis of young leaves. Symptoms of sulfur deficiency can be mistaken for those of nitrogen deficiency, although in the case of sulfur deficiency the younger leaves rather than the older leaves become chlorotic first because of limited translocation of sulfur among plant tissues. Prediction of sulfur levels in hop plants based on soil tests is generally problematic because the results are difficult to interpret and can be unreliable.

Zinc

The symptoms of zinc deficiency vary considerably depending on environmental conditions. Plants with mild symptoms

Fig. 135. Ashy-gray interveinal necrosis caused by potassium deficiency. (Courtesy C. B. Skotland)

Fig. 136. Weak growth of bines and lateral branches caused by zinc deficiency. Note extensive chlorosis of leaves above. (Courtesy C. B. Skotland)

have long shoots with very small, yellow leaves that are cupped upward and have deeply cut lobes with sharp marginal serrations. In severe cases, the serrations wither and leaves become brittle. Cones on affected plants are small and rattle when moved. Extreme cases can result in plant death. Zinc sulfate can be applied to the leaves to alleviate symptoms as soon as they appear, while soil amendment may be needed for the following growing season (Fig. 136). Zinc amendments also are reported to cause remission of symptoms associated with *Apple mosaic virus* and crinkle disease.

Selected References

Burgess, A. H. 1964. Hops: Botany, Cultivation and Utilization. World Crop Books, Interscience Publishers, New York.

Neve, R. A. 1991. Hops. Chapman and Hall, London

Rybáček, V., ed. 1991. Hop Production. Developments in Crop Science (Netherlands), vol. 16. Elsevier Science, Oxford.

(Prepared by W. F. Mahaffee, S. J. Pethybridge, and D. H. Gent)

Injuries Caused by Environmental Factors

Frost Damage

Temperatures at or near 0°C can cause damage to leaves and stems, particularly young, newly emerged tissues. Frost damage is most common in early to mid spring. Symptoms may include chlorosis or slight necrosis of leaves and shoot tips on new growth. Affected shoots may be stunted, and older leaves may have a rough, silvery appearance, which could be mistaken for downy mildew (Fig. 137). Plants generally recover from spring frost damage.

Hail Damage

Commercial hops are highly susceptible to hail damage. Growing tips of the main bines are usually broken and knocked off the string, which requires retraining. Leaves are usually torn and tattered so that little photosynthetic area remains (Fig. 138). If

Fig. 137. Interveinal bronzing of leaves caused by frost injury. (Courtesy D. H. Gent)

Fig. 138. Defoliation of plants caused by hail. (Courtesy E. Atkins)

Fig. 139. Nondescript browning of hop cones caused by cones rubbing against leaves and bines during wind events. (Courtesy W. F. Mahaffee)

flowering has occurred or is occurring, lateral branches are often broken and many of the flowers or cones are damaged or lost.

Wind Damage

Wind can damage hops in several ways. The most common is cone damage. Winds cause the cones to rub against abrasive leaves and bine stems, resulting in necrotic spots of various

Fig. 140. Salvaging hops from a collapsed trellis after high winds. (Courtesy D. Whitener)

Fig. 141. Leaf cupping caused by clopyralid. (Courtesy D. H. Gent)

sizes on the exposed portions of bracts and bracteoles (Fig. 139). This damage can be extensive and results in reduced cone quality. These necrotic areas can also serve as sites of infection for necrotrophic fungi (e.g., *Botrytis cinerea* Pers. and *Alternaria* spp.). High winds can cause lateral branches to break and even trellises to collapse (Fig. 140). In general, the latter occurs as a result of high winds and heavy rain just prior to harvest.

(Prepared by W. F. Mahaffee, S. J. Pethybridge, and D. H. Gent)

Chemical Injury

Improper use or drift of chemicals can cause temporary damage to or death of plants. Injury can result from the carryover of soil-active pesticides, excessive application rates, applications at an improper stage of plant development, insufficient cleaning of sprayer tanks, and incompatible mixtures of compounds. Contact herbicides (e.g., carfentrazone-ethyl and paraquat) may cause chlorotic or necrotic spots, leaf curling and twisting, and death of growing points. Systemic herbicides may cause mild to severe stunting, chlorosis or bleaching of leaves, epinasty, and deformation of leaves and cones. Symptoms caused by chemical injury often diminish over time, and affected plants recover in the current or subsequent season. Strict adherence to manufacturers' label recommendations, thorough cleaning of spray equipment, and avoidance of application of herbicides to growing shoot tips minimize injury.

Growth Regulators

Growth-regulating herbicides such 2,4-D, clopyralid (Fig. 141), quinclorac (Fig. 142), and triclopyr can damage hop when they are applied to or drift onto foliage and buds (Fig. 143) or from soil activity when high rates are applied. Injury by growth-regulator herbicides appears as mild to severe stunting of plants, distortion ("fiddleneck" appearance) and malformation of young leaves and shoots, leaf curling, and epinasty. Applications made later in the season can result in malformation of cones caused by curling of bracts and bracteoles.

Inhibitor of Pigment Biosynthesis

Norflurazon is a commonly applied soil herbicide that inhibits biosynthesis of carotenoid pigment. Symptoms of norflurazon injury appear initially in young growth and shoot tips as chlorotic or bleached areas along veins. Chlorosis and bleaching can later progress into interveinal areas (Fig. 144). Injury can result from application of high rates to coarse-textured soils, although symptoms are usually temporary.

Inhibitor of Amino Acid Biosynthesis

Glyphosate prevents biosynthesis of essential amino acids by inhibiting the enzyme 5-enolpyruvylshikimate-3-phosphate (EPSP) synthase. Injury to hop occurs when glyphosate comes in contact with foliage. Symptoms of glyphosate injury vary depending on when applications are made but may include interveinal chlorosis and bleaching, stunting, shortening of internodes, and narrow, malformed leaves. In addition, symptoms

Fig. 142. Severe epinasty, leaf curling, and distortion of cones caused by a midseason application of quinclorac. (Courtesy D. H. Gent)

Fig. 143. Chlorosis and distortion of leaves caused by a fall application of glyphosate, 2,4-D, and triclopyr for weedy *Rubus* spp. (Courtesy D. H. Gent)

Fig. 144. Chlorosis of leaf veins caused by an application of norflurazon. (Courtesy C. B. Skotland)

Fig. 145. Leaf chlorosis caused by a spring application of glyphosate. (Courtesy M. E. Nelson)

may appear on spring growth after fall application to foliage (Figs. 145 and 146).

Disruptors of Cell Membranes

Carfentrazone-ethyl. Carfentrazone-ethyl and its free acid metabolite inhibit the enzyme protoporphyringoen oxidase, resulting in generation of oxygen-free radicals that disrupt membranes and cause loss of cellular compartmentalization. Carfentrazone-ethyl may cause injury to hop when it contacts shoot meristems, foliage, and stems, which may result from direct application during stripping, drift, or carryover of carfentrazone-ethyl in spray tanks. Symptoms may include chlorotic to necrotic spots or necrosis of entire leaves, necrotic lesions on stems, photobleaching, stunting of plants, and death of developing lateral branches (Fig. 147). Certain *Fusarium* spp. may colonize stem lesions caused by carfentrazone-ethyl. The risk of crop injury increases when applications are made to well-watered yards or following several days of cloudy weather. Crop injury can be minimized by thoroughly cleaning spray tanks, directing applications of carfentrazone-ethyl to the lower 0.5 m of bines, and avoiding applications to newly trained bines.

Paraquat. Paraquat intercepts electrons in the light reaction of photosynthesis, resulting in the generation of oxygen-free radicals that react nonspecifically with lipid membranes and other molecules in the chloroplast. Treated plants rapidly lose chloroplast activity. Paraquat may cause injury to hop when it contacts shoot meristems, foliage, and stems, which may result from direct application during stripping, drift, or carryover in spray tanks. Symptoms may include chlorotic to necrotic spots on leaves, necrotic lesions on and cracking of stems, photobleaching, stunting of plants, and death of developing lateral branches and shoots if paraquat contacts the growing tip (Fig. 148). Certain *Fusarium* spp. may colonize stem lesions caused by paraquat. The risk of crop injury varies depending on temperature, the timing of applications, and cultivar. Crop injury can be minimized by thoroughly cleaning spray tanks, directing applications to the lower 0.5 m of bines, avoiding applications to newly trained bines, and avoiding applications made in the dark since this increases paraquat translocation.

Fig. 146. Leaf distortion and loss of pigment in leaves in spring caused by a fall application of glyphosate. (Courtesy D. H. Gent)

Fig. 147. Photobleaching of leaves and superficial necrosis of bines caused by carfentrazone-ethyl. (Courtesy D. H. Gent)

Fig. 148. Chlorosis and necrosis of shoots from a spring application of paraquat. (Courtesy D. H. Gent)

Fig. 149. Chlorosis caused by multiple applications of crop oil. (Courtesy D. H. Gent)

Fungicides

Crop oil. Repeated application of crop oils can reduce photosynthetic activity of leaves and cause a general chlorosis of leaves (Fig. 149). Crop oil injury can be more severe on plants stressed by other environmental conditions.

Sulfur. Application of sulfur when temperatures will be above 30°C can result in a blotchy yellowing of leaves. However, plants quickly recover with little or no permanent injury. Tank mixes of sulfur and crop oils can exacerbate injury, especially on young, succulent tissues.

Tebuconazole. Applications of tebuconazole to burrs and developing cones can lead to reductions in cone size and weight. Tebuconazole has also been implicated in delayed crop maturity. The effects of tebuconazole can be avoided if applications are restricted to pre-burr.

(Prepared by D. H. Gent, S. J. Pethybridge, and W. F. Mahaffee)

Heptachlor Wilt

Heptachlor is a cyclodiene insecticide that was used widely on many crops including potato, strawberry, and sugar beet. It was used extensively in the United States in 1955 and 1956 for control of strawberry root weevil on hop, and this led to severe die-out in treated hop yards. Decline was also reported in Alsace, France, in hop yards formerly treated with heptachlor. The use of heptachlor was discontinued on all crops in the United States in 1972. Heptachlor and its degradation product heptachlor epoxide in particular are very persistent in the soil and can still cause significant damage when hop yards are established in soil that was treated with this insecticide. Although it is believed to persist in most soil types, damaging levels of residues appear to be more persistent in sandy soils than in soils high in organic matter. Similar damage has been reported for the closely related pesticide chlordane. The technical chlordane used in commercial chlordane pesticide formulations was a mixture of 60–75% chlordane isomers and 10% heptachlor, with the remainder composed of many other compounds. Eventually heptachlor and heptachlor epoxide will degrade and no longer be a problem. Although cyclodiene pesticides can be detected at levels as low as 1–10 ng per gram of soil, hop seems to be sensitive to heptachlor and heptachlor epoxide at concentrations below detection limits. As a result, it is unknown how long the problem will persist.

Heptachlor-contaminated soil can result in extensive wilting and death of hop plants. While most factors that cause die-out result in more or less random distribution throughout the hop yard, heptachlor results in widespread and relatively uniform death of hop plants, often with distinct boundaries where applications ceased.

Symptoms of heptachlor wilt include a rough and corky epidermis or bark that cracks and exudes sap (Fig. 150). Plants exhibit decreased vigor and often die. Beginning at the outer margins, leaves become yellow and die, and bines often wilt and die while they are still fairly green. The bases of bines often enlarge and taper to a small diameter where they attach to the crown, breaking easily because they are brittle (Fig. 151). Similar basal swelling symptoms have been reported for Fusarium canker. Early crown symptoms occur in the xylem,

Fig. 150. Corky hop stem caused by heptachlor (bottom) and healthy stem (top). (Courtesy M. E. Nelson)

Fig. 151. Cracking hop bine caused by heptachlor. (Courtesy M. E. Nelson)

Fig. 152. Cross section of hop stems with various stages of rot caused by heptachlor. Top left healthy. (Courtesy M. E. Nelson)

where brown spotting develops. These spots expand, developing into rotting areas, and the entire crown eventually rots (Fig. 152). Crown rot symptoms are similar to those associated with *Pseudoperonospora humuli* (Miyabe & Takah.) G. W. Wilson, *Fusarium* spp., and *Phytophthora citricola* Sawada. Roots (as distinguished from the crown and rhizomes) often develop a black rot with very sharp margins between the healthy and rotted tissues (Fig. 153). When hops are planted in soil contaminated with heptachlor, the young plants initially grow normally

Fig. 153. Hop crowns with various stages or rotting caused by heptachlor. Top right, healthy. (Courtesy M. E. Nelson

but often cannot establish a root system and die during the summer. Increased infection by *Verticillium dahliae* Kleb. has been reported in soils treated with heptachlor.

It appears that some cultivars are more tolerant to heptachlor than others. For example, it was reported that Late Cluster cultivars appeared to be somewhat more tolerant than Early Cluster. Little has been reported regarding most currently grown cultivars, but it appears that many of the triploid cultivars released in the late 1980s and 1990s in the United States are fairly susceptible to heptachlor wilt.

Management

If heptachlor is present in the soil, options are limited. Economic production of hop is often impossible in heptachlor-treated yards. No treatments are known to alleviate the problem once hops are planted in contaminated soil; therefore, the only management options available occur prior to planting. Few site-specific pesticide records are available prior to 1972, but if any do exist, they should be examined for indications of heptachlor use. Before planting a new yard to hop, the top 25 cm of soil should be sampled and tested for heptachlor and heptachlor epoxide residues. If heptachlor is detected, hops should not be planted. However, a negative soil test is not a reliable indicator of the lack of potential heptachlor wilt problems because susceptible cultivars can be affected by heptachlor and heptachlor epoxide at concentrations below current detection limits. If soil tests are negative, test plots of the desired cultivar should be planted in the suspect soil and growth should be evaluated for at least 1 year prior to planting the entire yard.

Selected References

Klostermeyer, E. C., and Skotland, C. B. 1959. Pesticide chemicals as a factor in hop die out. Stn Circ. 362. Washington State University, Pullman.

Laouedj, A., Schenk, C., Pfohl-Leszkowicz, A., Keith, G., Schontz, D., Guillemaut, P., and Rether, B. 1995. Detection of DNA adducts in declining hop plants grown on fields formerly treated with heptachlor, a persistent insecticide. Environ. Pollut. 90:409-414.

McEwen, F. L., and Stephenson, G. R. 1979. The Use and Significance of Pesticides in the Environment. John Wiley and Sons, New York.

Skotland, C. B., and Romanko, R. R. 1987. The pesticide heptachlor and its influence on verticillium wilt of hops, *Humulus lupulus*. Can. J. Plant Pathol. 9:86.

(Prepared by M. E. Nelson)

Glossary

C—Celsius or centigrade
cm—centimeter (1 cm = 0.01 m = 0.3937 in.)
Da—dalton
g—gram (1 g = 0.03527 oz.)
h—hour
ha—hectare (1 ha = 2.471 acres)
kg—kilogram (1 kg = 2.205 lb)
m—meter (1 m = 39.37 in.)
mg—milligram (1 mg = 0.001 g)
ml—milliliter (1 ml = 0.001 liter)
mm—millimeter (1 mm = 0.001 m = 0.03937 in.)
µm—micrometer (1 µg = 10^{-6} m)
nm—nanometer (1 nm = 10^{-9} m)

abaxial—directed away from the stem of a plant; pertaining to the lower surface of a leaf (*see* adaxial)
abiotic—pertaining to the absence of life, as in diseases not caused by living organisms
abscission—the shedding of leaves or other plant parts as the result of physical weakness in a specialized layer of cells (the abscission layer) that develops at the base of the structure
acid—having a pH less than 7
acute—pertaining to symptoms that develop suddenly
adaxial—directed toward the stem of a plant; pertaining to the upper surface of a leaf (*see* abaxial)
aerial—occurring in the air
aerobic—living only in the presence of oxygen
agar—jellylike material derived from algae and used to solidify liquid culture media; a medium containing agar
alkaline—having basic (nonacidic) properties; having a pH greater than 7
alternate host—one of two kinds of plant on which a parasitic fungus (e.g., rust) must develop to complete its life cycle
amino acid—organic nitrogenous acids from which protein molecules are constructed
anaerobic—living in the absence of oxygen
anamorph (adj. anamorphic; syn. imperfect state)—the asexual form in the life cycle of a fungus, when asexual spores (such as conidia) or no spores are produced (*see* teleomorph)
annual—a plant that completes its life cycle and dies within one year (*see* perennial)
antagonist—an organism or substance that limits or counteracts the action of another
anterior—toward the front or head
anther—pollen-bearing portion of a flower
antibiotic—a chemical compound produced by one microorganism that inhibits growth or kills other living organisms
antigen—any foreign chemical (normally a protein) that induces antibody formation in warm-blooded animals
antiserum (pl. antisera)—blood serum containing antibodies
apex (pl. apices; adj. apical)—tip of a root or shoot, containing the apical meristem
aphid—small, sucking insect of the family Aphididae (order Homoptera) that produces honeydew and injures plants when in large populations
apothecium (pl. apothecia)—open cuplike or saucerlike ascus-bearing fungal fruiting body (ascocarp), often supported on a stalk
appressed (syn. adpressed)—closely flattened down or pressed against a surface
appressorium (pl. appressoria)—swollen, flattened portion of a fungal filament that adheres to the surface of a higher plant, providing anchorage for invasion by a fungus
apterous—wingless
arachnoid—similar to a spider's web
arthropod—member of the phylum Arthropoda, which consists of animals with articulated bodies and limbs and which includes insects, arachnids, and crustaceans
ascocarp (syn. ascoma)—sexual fruiting body of an ascomycetous fungus that produces asci and ascospores; e.g., apothecium, ascostroma, cleistothecium, perithecium, and pseudothecium
ascospore—sexual spore borne in an ascus
ascus (pl. asci)—saclike structure containing ascospores (typically eight) and usually borne in a fungal fruiting body
aseptate—having no cross walls; lacking septa; nonseptate; coenocytic
asexual—vegetative; without sex organs, gametes, or sexual spores; the imperfect or anamorphic stage of a fungus
asexual reproduction—any type of reproduction not involving the union of gametes and meiosis
attenuate (v. or adj.)—to narrow; to weaken or decrease in virulence or pathogenicity
avirulent (syn. nonpathogenic)—unable to cause disease (*see* virulent)
axenic—culture in the absence of living bacteria or other organisms; pure culture
axil—the angle formed by a leaf petiole and the stem
axillary—pertaining to or placed within an axil

bacterium (pl. bacteria)—a prokaryotic, microscopic, single-celled organism with a cell wall that increases by binary fission
basidiocarp (syn. basidioma)—sexual fruiting body of a basidiomycetous fungus
basidiospore—haploid (1N) sexual spore produced on a basidium
basidium (pl. basidia; adj. basidial)—a structure on which haploid basidiospores are produced externally
biguttulate—having two droplets or globules of oil
binary fission—a type of asexual reproduction in which two cells, usually of similar size and shape, are formed by the growth and division of one cell
bine—flexible twining or climbing stem of certain plants
biological control (syn. biocontrol)—exploitation by humans of the natural competition, parasitism, or antagonism of organisms for management of pests and pathogens
biotic—relating to life, as in disease caused by living organisms
biotroph (syn. obligate parasite)—an organism that can live and multiply only on another living organism (*see* necrotroph)
biotype—a subdivision of a species, subspecies, or race based on some identifiable physiological trait such as a specific virulence pattern
biovar—a group of microorganisms, usually bacteria, that have identical genetic but different biochemical or physiological characters
blade—the flat portion of a grass leaf above the sheath
blastospore—asexual reproductive spore that originates by budding
blight—sudden, severe, and extensive spotting, discoloration, wilting, or destruction of leaves, flowers, stems, or entire plants
blotch—irregularly shaped, usually superficial spot or blot
bract—a reduced leaf associated with a flower or inflorescence; modified leaf from the axil of which a flower arises
bracteole—organ resembling a leaf or scale that arises from a branch of a flower cluster and develops above a bract

broadcast application—fertilizer application by spreading or scattering on the soil surface

caducous—sloughing off or falling away early
calcareous—rich in calcium carbonate (lime)
callose—a plant cell wall polysaccharide found in phloem sieve plates, wounded tissue, and certain specialized cells
canker—a plant disease characterized (in woody plants) by the death of cambium tissue and loss or malformation of bark, or (in nonwoody plants) by the formation of sharply delineated, dry, necrotic, localized lesions on the stem; lesion caused by such a disease, particularly in woody plants
canopy—the expanded leafy top of a plant or plants
capsid (syn. coat protein)—the protective layer of protein surrounding the nucleic acid core of a virus; the protein molecules which make up this layer
capsomer—the basic structural unit of a virus particle
carbohydrate—any of various chemical compounds composed of carbon, hydrogen, and oxygen, such as sugars, starches, and cellulose
causal agent—organism or agent that produces a given disease
chasmothecium (pl. chasmothecia)—sexual fruiting body of powdery mildew fungi (*see* cleistothecium)
chemotactic—pertaining to the movement or orientation of an organism or cell along a chemical concentration gradient either toward or away from the chemical stimulus
chlamydospore—thick-walled or double-walled asexual resting spore formed from hyphal cells (terminal or intercalary) or by transformation of conidial cells that can function as an overwintering stage
chloroplast—disklike structure containing chlorophyll in which photosynthesis occurs in green plants
chlorosis (adj. chlorotic)—failure of chlorophyll development, caused by disease or a nutritional disturbance; fading of green plant color to light green, yellow, or white
chromosome—the structure that contains the genes of an organism; in eukaryotes, chromosomes are in the nucleus and can be visualized with an optical microscope as threads or rods during meiosis and mitosis; in bacteria, the chromosome is usually a single circle of DNA that cannot be visualized with an optical microscope
cirrhus (pl. cirrhi)—a curled, tendril-like mass of exuded spores, held together by a slimy matrix; also cirrus
clavate (syn. claviform)—club shaped
cleistothecium (pl. cleistothecia)—a spherical ascocarp that is closed at maturity (*see* chasmothecium)
coalesce—grow together into one body or spot
coenocytic—having multiple nuclei embedded in cytoplasm without cross walls; nonseptate
colonization—establishment and ramification of a pathogen within a host plant
colonize—to infect and ramify through plant tissue with the growth of a pathogen
colony—growth of a microorganism in mass, especially as a pure culture
conidiogenous—producing and bearing conidia
conidiophore—simple or branched hypha on which conidia are produced
conidium (pl. conidia)—an asexual, nonmotile fungal spore that develops externally or is liberated from the cell that formed it
cordate—heart shaped
cortex (adj. cortical)—region of parenchyma tissue between the epidermis and the phloem in stems and roots; region beneath the rind of a sclerotium
cotyledon—seed leaf, one in monocots and two in dicots; primary embryonic leaf within the seed in which nutrients for the new plant are stored
crepuscular—active in low light
cross-protection—the process whereby a normally susceptible host is infected with a less virulent pathogen (usually a virus) and thereby becomes resistant to infection by a second, usually related, more virulent pathogen
crown—upper dome of a tree, bearing leaves, flowers, and fruits; junction of root and stem of a plant, usually at the soil line; in grafted woody plants, the rootstock portion of the plant near the soil surface
cultivar (abbr. cv.; syn. variety)—a plant type within a species, resulting from deliberate genetic manipulation, which has recognizable characteristics (color, shape of flowers, fruits, seeds, and height or form)
cultural practices—the manner in which plants are grown, such as application of nutrients, irrigation practices, type of cultivation; may be used for disease management
culture—growth and propagation of microorganisms on nutrient media; growth and propagation of living plants
cuticle (adj. cuticular)—noncellular outer layer of an insect or a nematode; water-repellent, waxy layer of epidermal cells of plant parts, such as leaves, stems, and fruit
cv.—*see* cultivar
cylindrical—shaped like a cylinder
cyst—in fungi, a resting structure in a protective membrane or shell-like enclosure; in nematode females, the egg-laden carcass of a female nematode; in bacteria, a specialized type of bacterial cell enclosed in a thick wall, often dormant and resistant to environmental conditions

decussate—to cross in the form of an X; intersect
defoliation—loss of leaves from a plant, whether normal or premature
desiccate—to dry out
determinate—pertaining to a sporogenous structure that does not continue developing after a spore has detached from its point of origin
deutonymph—second of up to three nymphal stages found in mites
diagnostic (n. diagnosis)—pertaining to a distinguishing characteristic important for the identification of a disease or other condition
diapause—temporary period of low metabolic activity in an animal or insect in which development is suspended
dichotomous—branching, often successively, into two more or less equal arms
dicot (syn. dicotyledon)—plant with two cotyledons or seed leaves (*see* monocot)
dieback (v. die back)—progressive death of shoots, leaves, or roots, beginning at the tips
dimorphic—having two distinct shapes or forms
dioecious—having the male and female organs in separate and distinct individuals; having separate sexes (*see* monoecious)
diploid—having two complete sets of chromosomes (2N chromosomes) (*see* haploid)
disease—abnormal functioning of an organism
disease cycle—succession of all events and interactions among the host, parasite, and environment that occur in a disease, from initial infection of the plant by a causal agent, through pathogenesis, to overseasoning, until another infection occurs
disinfect—to eliminate a pathogen from infected plant tissues
dispersal (syn. dissemination)—spread of infectious material (inoculum) from diseased plants to healthy plants
dissemination (syn. dispersal)—spread of infectious material (inoculum) from diseased plants to healthy plants
distal—far from the point of attachment or origin; in a direction away from main body
DNA (abbr. for deoxyribonucleic acid)—the double-stranded, helical molecule that contains genetic code information; each repeating unit, or nucleotide, is composed of deoxyribose (a sugar), a phosphate group, and a purine (adenine or guanine) or a pyrimidine (thymine or cytosine) base
doliiform—barrel shaped
dominant—describes a phenotypic trait that is expressed in hybrid progeny of diploid organisms even when contributed by only one of the parents
dormancy (adj. dormant)—a condition of suspended growth and reduced metabolism of an organism, generally induced by internal factors or environmental conditions as a mechanism of survival
dormant—resting; living in a state of reduced physiological activity
dorsal—pertaining to the back or top (as opposed to ventral)
dwarfing—underdevelopment of a plant or plant organs, which may be caused by disease, inadequate nutrition, or unfavorable environmental conditions

eclose—to emerge from the egg or pupal cuticle
ecology—the study of the interactions that occur between individual organisms, groups of organisms, and organisms and their environment
egg mass—group of eggs held together by a gelatinous matrix
electrophoresis—separation of molecules according to their electric charge

ELISA—*see* enzyme-linked immunosorbent assay
ellipsoid—elliptical in plane section
elytra—hardened or rigid forewing, especially of members of the order Coleoptera
emergence—growth of the seedling shoot through the surface of the soil
enation—abnormal outgrowth from the surface of a stem or leaf
encapsidate—to cover virus nucleic acid with a protein coat
encyst—to form a cyst or protective covering
endemic—native to a particular place; pertaining to a low and steady level of natural disease occurrence
endogenous—arising from the inside (*see* exogenous)
endoparasite—parasitic organism that lives and feeds from inside its host
entomopathogenic—causing disease to an insect
enzyme—protein that catalyzes a specific biochemical reaction
enzyme-linked immunosorbent assay (acronym ELISA)—a serological test in which the sensitivity of the reaction is increased by attaching an enzyme that produces a colored product to one of the reactants
epidemic—an increase of disease in a population; a general and serious outbreak of disease
epidemiology (adj. epidemiological)—the study of factors influencing the initiation, development, and spread of infectious disease; the study of disease in populations of plants
epidermis (adj. epidermal)—surface layer of cells of leaves and other plant parts
epinasty—abnormal, downward curling of a leaf, leaf part, or stem
epiphyte—organism growing on a plant surface, but not as a parasite
eradication—control of plant disease by eliminating the pathogen after it is established or by eliminating the plants that carry the pathogen
etiology—the study of the causes of diseases
eukaryote—an organism containing a membrane-bound nucleus and other organelles, i.e., all higher plants, animals, fungi, and protists (*see* prokaryote)
exogenous—originating from the outside (*see* endogenous)
explant—to transfer part of a plant from the original and place it in a culture medium for growth
extracellular—outside a cell
exudates—liquid excreted or discharged from diseased tissues, from roots and leaves, or by fungi

fallow—cultivated land kept free from a crop or weeds during the normal growing season
fastidious—in reference to prokaryotic organisms, having special growth and nutritional requirements
filament (adj. filamentous)—a thin, flexible, threadlike structure
filamentous (syn. filiform)—threadlike
filiform—long, needlelike
flagellate—having one or more flagella
flagellum (pl. flagella)—hairlike, whiplike, or tinsel-like appendage of a motile cell, bacterium, or zoospore that provides locomotion
fleck—minute, discolored spot in green tissue
flexuous—having turns or windings; capable of bending
foliar—pertaining to leaves
fumigant (v. fumigate)—a gas or volatile substance used to kill or inhibit the growth of microorganisms or other pests
fundatrigenium (pl. fundatrigenia)—wingless aphid produced parthenogenetically by live birth
fundatrix—female aphid emerging from an overwintered egg that initiates colonies in spring
fungicide (adj. fungicidal)—chemical or physical agent that kills or inhibits the growth of fungi

gall—abnormal swelling or localized outgrowth, often roughly spherical, produced by a plant as a result of attack by a fungus, bacterium, nematode, insect, or other organism
gene—unit within an organism controlling heritable characteristics; genes are organized on chromosomes
genetic—relating to heredity; referring to heritable characteristics
geniculate—bent, like a knee
genome—the complete genetic information of an organism or virus
genotype—genetic constitution of an individual or group; class or group of individuals sharing a specific genetic makeup
genus (pl. genera)—a taxonomic category that includes a group of closely related (structurally or phylogenetically) species; the genus or generic name is the first name in a Latin binomial
germ tube—hypha resulting from an outgrowth of the spore wall and cytoplasm after germination
germinate (n. germination)—begin growth (as of a seed, spore, sclerotium, or other reproductive body)
germplasm—bearer of heredity material, often loosely applied to cultivars and breeding lines
girdle—to circle and cut through a stem or the bark and outer few rings of wood, disrupting the phloem and xylem
globose—nearly spherical
graft transmission—transmission of a pathogen from one host plant to another through fusion of living tissue from the diseased host with living tissue of a healthy host
gram-negative—bacteria staining red or pink in the Gram staining procedure after treatment with Gram's stain
gram-positive—bacteria staining violet or purple in the Gram staining procedure after treatment with Gram's stain
guttulate—containing droplets or globules or oil
gymnosperm—a vascular plant having seeds that are not enclosed in an ovary

haploid—having a single complete set of chromosomes (*see* diploid)
haustorium (pl. haustoria)—specialized branch of a parasite formed inside host cells to absorb nutrients
herbaceous—describing primary, soft, nonwoody tissue, as a plant or plant part; having the characteristics of an herb
herbicide—a chemical used for killing plants or inhibiting plant growth, e.g., a weed or grass killer
heterothallism (adj. heterothallic)—condition in which sexual reproduction can occur only in the presence of genetically different mycelia (*see* homothallism)
holocyclic—having a complete life cycle that includes sexual and parthenogenetic reproduction
homothallism (adj. homothallic)—condition in which sexual reproduction occurs with a single thallus; self-fertile (*see* heterothallism)
honeydew—sugary ooze or exudate, often from aphids, and a characteristic symptom of ergot
host—living plant attacked by or harboring a living parasite and from which the invader obtains part or all of its nourishment
host plant—living plant attacked by or harboring a parasite or pathogen and from which the invader obtains part or all of its nourishment
host range—the range of plants on which an organism, particularly a parasite, feeds
hyaline—transparent or nearly so; translucent; often used in the sense of colorless
hybrid (v. hybridize)—offspring of two individuals of different genotypes
hypha (pl. hyphae; adj. hyphal)—single, tubular filament of a fungal thallus or mycelium; the basic structural unit of a fungus

immune—cannot be infected by a given pathogen
immunosorbent electron microscopy—detection of virus particles through reaction with specific antibodies which are visualized using electromagnetic lenses to focus a beam of electrons
imperfect state (syn. anamorph)—the asexual form in the life cycle of a fungus, when asexual spores (such as conidia) or no spores are produced
in vitro—in glass, on artificial media, or in an artificial environment; outside the host
in vivo—within a living organism
indicator plant—plant that reacts to a pathogen or an environmental factor with specific symptoms, used to detect or identify the pathogen or determine the effects of the environmental factor
infect—process in which an organism enters, invades, or penetrates and establishes a parasitic relationship with a host plant
infection court—a site in or on a host plant where infection can occur
infectious—pertaining to disease, capable of spreading from plant to plant
infest (n. infestation)—to attack as a pest (used especially of insects and nematodes); to contaminate, as with microorganisms; to be present in large numbers
inflorescence—flower or flower cluster

initial inoculum (syn. primary inoculum)—inoculum, usually from an overwintering source, that initiates disease in the field, as opposed to inoculum that spreads disease during the season
injury—damage caused by transitory interaction with an agent such as an insect, chemical, or unfavorable environmental condition
inoculate (n. inoculation)—to place inoculum in an infection court; to insert a pathogen into healthy tissue
inoculum (pl. inocula)—pathogen or its parts, capable of causing infection when transferred to a favorable location
inoculum density—a measure of the number of propagules of a pathogenic organism per unit area or volume
instar—arthropod stage between two successive molts
intercellular—between or among cells
internode (adj. internodal)—the portion of a stem between two successive nodes
interveinal—between (leaf) veins
intracellular—through or within cells
isolate (n.)—a culture or subpopulation of a microorganism separated from its parent population and maintained in some sort of controlled circumstance; (v.) to remove from soil or host material and grow in pure culture
isometric—usually used for virus particles to describe those that are icosahedral in structure and appear approximately round

juvenile—an immature form that appears similar to but usually smaller than the adult and is not sexually mature (e.g., insects with gradual metamorphosis, nematodes)

kairomone—substance produced by an organism that attracts other species
knot—gall; a localized abnormal swelling

larva (pl. larvae)—immature stage of certain animals (especially insects) that undergo complete metamorphosis; the wormlike or caterpillar stage of the life cycle of such insects
latent—present but not manifested or visible, as a symptomless infection by a pathogen
latent infection—infection unaccompanied by visible symptoms
latent period—the time between infection and the appearance of symptoms or the production of new inoculum (sometimes synonymous with incubation period); the time after a vector has acquired a pathogen and before it can be transmitted
leaf spot—a plant disease lesion typically restricted in development in the leaf after reaching a characteristic size
lesion—localized diseased area or wound
life cycle—the cyclical stages in the growth and development of an organism
local lesion—small, restricted lesion, often the characteristic reaction of differential cultivars to specific pathogens, especially in response to mechanical inoculation with a virus

macroconidium (pl. macroconidia)—the larger of two kinds of conidia formed by certain fungi (*see* microconidium)
mating types—compatible strains, usually designated + and – or A and B, necessary for sexual reproduction in heterothallic fungi
mechanical injury—injury of a plant part by abrasion, mutilation, or wounding
mechanical transmission—spread or introduction of inoculum to an infection court (wounding) by human manipulation, accompanied by physical disruption of host tissues
medium (pl. media)—a mixture of organic and inorganic chemical compounds and water that provides the nutrients needed for the growth of a microorganism in vitro; for higher plants, a mixture of fertilizers and other components in which a plant is growing
meristem (adj. meristematic)—plant tissue characterized by frequent cell division, producing cells that become differentiated into specialized tissues
meristem culture—aseptic culture of a plant or plant part from a portion of the meristem
metabolite—any chemical participating in metabolism; a nutrient
microclimate—weather conditions on a small scale, e.g., at the surface of the plant or within a crop
microconidium (pl. microconidia)—the smaller of two kinds of conidia formed by certain fungi (*see* macroconidium)
microorganism (syn. microbe)—an organism of such small size that it can only be seen as an individual organism with the aid of a microscope
microsclerotium—microscopic, dense aggregate of darkly pigmented, thick-walled hyphal cells
midrib—central, thickened vein of a leaf
migratory—moving from place to place on a plant or from plant to plant when feeding (*see* sedentary)
mildew—thin coating of mycelial growth and spores on the surfaces of infected plant parts
mold—any microfungus with conspicuous, profuse, or woolly superficial growth (mycelium or spore masses) on various substrates; especially an economically important saprobe; molds commonly grow on damp or decaying matter and on the surface of plant tissues
molecular weight—sum of the atomic weights of all the atoms constituting a molecule
molt—to shed a cuticle or body encasement during a phase of growth
monocot (syn. monocotyledon)—plant with one cotyledon, such as grasses, grain crops, and corn (*see* dicot)
monoecious—having male and female reproductive organs on a single individual (*see* dioecious)
morphology (adj. morphological)—the study of the form of organisms; form and structure of organisms
mosaic—disease symptom characterized by nonuniform coloration, with intermingled normal, light green, and yellowish patches, usually caused by a virus; often used interchangeably with mottle
motile—capable of self-propulsion by means of flagella, cilia, or amoeboid movement
mottle—disease symptom composed of light and dark areas in an irregular pattern, usually caused by a virus; often used interchangeably with mosaic
multinucleate—having more than one nucleus per cell
mycelium (pl. mycelia; adj. mycelial)—mass of hyphae constituting the body (thallus) of a fungus

necrosis (adj. necrotic)—death of cells or tissue, usually accompanied by black or brown darkening
necrotroph—a parasite that typically kills and obtains its energy from dead host cells (*see* biotroph)
nematicide—agent, usually a chemical, that kills nematodes
nematode—nonsegmented roundworm (animal), parasitic on plants or animals, or free living in soil or water
node (adj. nodal)—enlarged portion of a shoot at which leaves or buds arise
nonpersistent transmission (syn. stylet-borne transmission)—a type of virus transmission in which the virus is acquired and transmitted by the vector after short feeding times and is retained by the vector for only a short period of time
nonseptate—without cross walls; coenocytic
nucleocapsid—the genome and protein coat of a virus
nucleotide—a subunit of a nucleic acid, either DNA or RNA
nymph—the juvenile stage in the life cycle of an insect, with incomplete metamorphosis but superficially resembling the adult

obligate parasite (syn. biotroph)—organism that can grow only as a parasite in association with its host plant and cannot be grown in artificial culture media
obtuse—rounded or blunt; greater than a right angle
Oomycetes (adj. oomycetous)—class of fungal-like organisms typically with nonseptate mycelium, asexual sporangia and zoospores, and sexual oospores
oospore—thick-walled, sexually derived resting spore of fungi in the class Oomycetes
ostiole (adj. ostiolate)—pore; opening in the papilla or neck of a perithecium, pseudothecium, or pycnidium through which spores are released
ovate—having a shape like the longitudinal section of an egg
overwinter—to survive or persist through the winter period
oviposit—to deposit or lay eggs with an ovipositor
ovule—enclosed structure that, after fertilization, becomes a seed; egg contained within an ovary

palmate—having four or more lobes or leaflets radiating from a single point

panicle—a loose, diversely branching flower cluster
papilla (pl. papillae; adj. papillate)—small, blunt projection
parasite (adj. parasitic)—organism that lives in intimate association with another organism on which it depends for its nutrition; not necessarily a pathogen
parasitoid—organism that spends a significant portion of its life attached to or within a host organism
parenchyma (adj. parenchymatous)—soft tissue of living plant cells with undifferentiated, thin, cellulose walls
parthenogenesis (adj. parthenogenetic)—reproduction by the development of an unfertilized egg
pathogen (adj. pathogenic)—a disease-producing organism or agent
pathogenicity—ability to cause disease
pathotype—a subdivision of a pathogen species characterized by its pattern of virulence or avirulence to a series of differential host varieties
PCR—*see* polymerase chain reaction
pectolytic—capable of hydrolysis of pectic substances
penetration—initial invasion of a host by a pathogen
perennate—to survive from one growing season to the next
perennial—something that occurs year after year; plant that survives for several to many years (*see* annual)
perithecium (pl. perithecia)—flask-shaped or subglobose, thin-walled fungus fruiting body (ascocarp) containing asci and ascospores; spores are expelled or released through a pore (ostiole) at the apex
peritrichous—having hairs or flagella distributed over the whole surface
persistent transmission (syn. circulative transmission)—a type of virus transmission in which the virus is acquired and transmitted by the vector after relatively long feeding times and remains transmissible for a prolonged period while in association with its vector
pest—any organism that damages plants or plant products
pesticide—a chemical used to control pests
petiole—stalk portion of a leaf
pH—negative logarithm of the effective hydrogen ion concentration; a measure of acidity (pH 7 is neutral; values less than pH 7, acidic; greater than pH 7, alkaline)
pheromone—a chemical compound produced by an animal or arthropod that influences the behavior and development of other members of the same species
phialide—end cell of a conidiophore with one or more open ends through which a basipetal succession of conidia develops
phloem—food-conducting, food-storing tissue in the vascular system of roots, stems, and leaves
photosynthate—product of photosynthesis
photosynthesis—manufacture of carbohydrates from carbon dioxide and water in the presence of chlorophyll(s), using light energy and releasing oxygen
phyllody—change of floral organs to leaflike structures
phytoplasma (syn. mycoplasmalike organism, MLO)—plant-parasitic pleomorphic mollicute (prokaryote with no cell wall) found in phloem tissue; cannot yet be grown on artificial nutrient media
phytosanitary—pertaining to the health of plants
phytotoxic—harmful to plants (usually used to describe chemicals)
pigment—a colored compound, such as chlorophyll, in the cells of plants or fungi
pith—parenchymatous tissue occupying the center of the stem
plant pathology (syn. phytopathology)—the study of plant diseases
pleomorphic—able to assume various shapes (and perhaps sizes); occurrence of more than one distinct form in the life cycle of an organism
pollen—male sex cells produced by anthers of flowering plants or the cones of seed plants
pollination—transfer of pollen from anther to stigma or from a staminate cone to an ovulate cone
polymerase chain reaction (acronym PCR)—a technique to amplify DNA
polyphagia (adj. polyphagous)—feeding on many different types of food
positive sense RNA—RNA that can serve directly as messenger RNA
predispose (n. predisposition)—to make prone to infection and disease
primary inoculum (syn. initial inoculum)—inoculum, usually from an overwintering source, that initiates disease in the field, as opposed to inoculum that spreads disease during the season
prokaryote—an organism without internal membrane-bound organelles, lacking a distinct nucleus, such as bacteria and mollicutes (*see* eukaryote)
prolegs—abdominal limbs found in certain insects
propagule—any part of an organism capable of independent growth
protectant—agent, usually a chemical, applied to a plant surface in advance of a pathogen to prevent infection
protein—a nitrogen-containing organic compound composed of units called amino acids
protonymph—first of up to three nymphal stages found in mites
protoplasm—living contents of a cell
pteridophyte—spore bearing plants such as club mosses, horsetails, and ferns
pubescence—a covering of down or fine short hair
pupate (n. pupa)—to develop from an arthropod larva to an adult
pycnidium (pl. pycnidia)—asexual, globose or flask-shaped fruiting body of certain imperfect fungi producing conidia

quarantine—legislative control of the transport of plants or plant parts to prevent the spread of pests or pathogens
quiescent—dormant or inactive

race—subgroup or biotype within a species or variety, distinguished from other races by virulence, symptom expression, or host range, but not by morphology
rachis—elongated main axis of an inflorescence
resistant (n. resistance)—possessing properties that prevent or impede disease development (*see* susceptible)
resting spore—spore, often thick-walled, that can remain alive in a dormant state for some time, later germinating and capable of initiating infection
reverse-transcriptase polymerase chain reaction (acronym RT-PCR)—a technique used to make and amplify complementary DNA from a piece of RNA
rhizome—a mostly horizontal, jointed, fleshy, often elongated, usually underground stem
rhizomorph—macroscopic ropelike strand of compacted tissue formed by certain fungi
rhizosphere—microenvironment in the soil, immediately around plant roots
ringspot—disease symptom characterized by yellowish or necrotic rings enclosing green tissue, as in some plant diseases caused by viruses
RNA (abbr. for ribonucleic acid)—several nucleic acids composed of repeating units of ribose (a sugar), a phosphate group, and a purine (adenine or guanine) or a pyrimidine (uracil or cytosine) base; transcribed from DNA and involved in translation to proteins
root cap—a group of cells on a root that protects the growing tip
rot—softening, discoloration, and often disintegration of plant tissue as a result of fungal or bacterial infection
rotation—growth of different kinds of crops in succession in the same field
RT-PCR—*see* reverse-transcriptase polymerase chain reaction

sanitation—destruction or removal of infected and infested plants or plant parts; decontamination of tools, equipment, containers, work space, hands, etc.
saprophyte (adj. saprophytic; syn. saprobe)—organism that obtains nourishment from nonliving organic matter
satellite nucleic acid—subviral agent composed of nucleic acid molecules that depends on co-infection of a host cell with a helper virus for their productive multiplication
sclerotium (pl. sclerotia)—a vegetative resting body of a fungus, composed of a compact mass of hyphae with or without host tissue, usually with a darkened rind
scorch—any symptom that suggests the action of flame or fire on the affected part, often seen at the margins of leaves
secondary infection—infection resulting from the spread of infectious material produced after a primary infection or from a secondary infection without an intervening inactive period
secondary inoculum—inoculum produced by infections that took place during the same growing season
secondary organism—organism that multiplies in already diseased tissue but is not the primary pathogen

sedentary—remaining in a fixed location (*see* migratory)
seed—ripened ovule consisting of an embryo and stored food enclosed by a seed coat
semipapillate—having very small, blunt projections
senesce (adj. senescent, n. senescence)—to decline, as with maturation, age, or disease stress
septate—with cross walls; having septa
septum (pl. septa, adj. septate)—dividing wall; in fungi, cross wall
serology (adj. serological)—a method using the specificity of the antigen-antibody reaction for the detection and identification of antigenic substances and the organisms that carry them
shot hole—symptom in which small lesions fall out of leaves, giving the leaf the appearance of being hit by buckshot
sieve element (syn. sieve tube element)—a tube-shaped living cell in the phloem functioning in the transport of dissolved organic substances in the plant
sign—indication of disease from direct observation of a pathogen or its parts (*see* symptom)
siphunculus (pl. siphunculi)—one of a pair of dorsal tubes on the abdomen of aphids that secrete alarm pheromones
soilborne—carried on or beneath the soil surface
sooty mold—black, nonparasitic, superficial fungal growth on honeydew produced by aphids and other phloem-feeding insects
sp. (abbr. for species; pl. spp.)—a genus name followed by sp. means that the particular species is undetermined; spp. after a genus name means that several species are being referred to
sporangiophore—sporangium-bearing body of a fungus
sporangium (pl. sporangia)—saclike fungal structure in which the entire contents are converted into an indefinite number of asexual spores
spore—reproductive structure of fungi and some other organisms, containing one or more cells; a bacterial cell modified to survive an adverse environment
sporodochium (pl. sporodochia)—superficial, cushion-shaped asexual fruiting body consisting of a cluster of conidiophores
sporulate—to produce spores
sporulating—producing and often liberating spores
spot—a symptom of disease characterized by a limited necrotic area, as on leaves, flowers, and stems
stamen (adj. staminal)—male structure of a flower, composed of a pollen-bearing anther and a filament, or stalk
stele—central cylinder of vascular tissue (especially in roots)
stellate—having the shape of a star
sterile—unable to reproduce sexually; to be free of living microorganisms
stigma—portion of a flower that receives pollen and on which the pollen germinates
stippling—series of small dots or speckles in which chlorophyll is absent
stipule—small leaflike appendage at the base of a leaf petiole, usually occurring in pairs
stoma (pl. stomata; adj. stomatal; also stomate)—structure composed of two guard cells and the opening between them in the epidermis of a leaf or stem, functioning in gas exchange
strain—a distinct form of an organism or virus within a species, differing from other forms of the species biologically, physically, or chemically
strig—main axis of the female hop inflorescence
strobile—a reproductive structure characterized by overlapping scalelike part
stunting—reduction in height of a vertical axis resulting from a progressive reduction in the length of successive internodes or a decrease in their number
style—slender part of many pistils located between the stigma and the ovary and through which the pollen tube grows
stylet—stiff, slender, hollow feeding organ of plant-parasitic nematodes or sap-sucking insects, such as aphids or leafhoppers
subhyaline—nearly translucent and colorless
substrate—the substance on which an organism lives or from which it obtains nutrients; chemical substance acted upon, often by an enzyme
susceptible (n. susceptibility)—prone to develop disease when infected by a pathogen (*see* resistant)
sympodium (pl. sympodia; adj. sympodial)—conidiophore which appears simple, but terminates and branches repeatedly
symptom—indication of disease by reaction of the host, e.g., canker, leaf spot, and wilt (*see* sign)
symptomless carrier—a plant that, although infected with a pathogen (usually a virus), produces no obvious symptoms
syncytium (pl. syncytia)—a multinucleate structure in root tissue formed by dissolution of common cell walls induced by secretions of certain sedentary plant-parasitic nematodes, e.g., cyst nematodes
synergism (adj. synergistic)—greater than additive effect of interacting factors
synomone—chemical signal released by members of one species that affects the behavior of another species and benefits both parties
systemic—pertaining to a disease in which the pathogen (or a single infection) spreads generally throughout the plant; pertaining to chemicals that spread internally through the plant

taxonomy (adj. taxonomic)—the science dealing with naming and classifying organisms
teleomorph—the sexual form in the life cycle of a fungus (*see* anamorph)
tendril—threadlike, leafless organ of climbing plants that attaches itself to or twines round some other body
tetraploid—having a chromosome number that is four times the basic or haploid number
thermotherapy—use of heat to reduce or eliminate pathogens in plant tissue; often used on plants prior to meristem culture to produce pathogen-free plants
tissue—group of cells, usually of similar structure, that perform the same or related functions
titer—measure of the amount or concentration of a substance in a solution
tolerance (adj. tolerant)—ability of a plant to endure an infectious or noninfectious disease, adverse conditions, or chemical injury without serious damage or yield loss; of pesticides, the amount of chemical reside legally permitted on an agricultural product entering commercial channels and usually measured in parts per million (ppm)
toxicity—capacity of a substance to interfere with the vital processes of an organism
toxin—poisonous substance of biological origin
translocation—movement of water, nutrients, chemicals, or food materials within a plant
translucent—so clear that light may pass through
transmit (n. transmission)—to spread or transfer, as in spreading an infectious pathogen from plant to plant or from one plant generation to another
transpiration—water loss by evaporation from leaf surfaces and through stomata
trichome—a plant epidermal hair, of which several types exist
tricussate—whorls of three leaves, each alternating with leaves at the nodes above and below
triploid—having a chromosome number that is three times the basic or haploid number
tuber—an underground stem adapted for storage, typically produced at the end of a stolon
tubercle—rounded protuberance on a plant or insect, particularly caterpillars
tumor—abnormal swelling or localized outgrowth, often roughly spherical, produced by a plant as a result of attack by a fungus, bacterium, nematode, insect, or other organism

unicellular—one celled

variegation—pattern of two or more colors in a plant part, as in a green and white leaf
variety (syn. cultivar; adj. varietal)—a plant type within a species, resulting from deliberate manipulation, which has recognizable characteristics (color, shape of flowers, fruits, seeds, height and form)
vascular—pertaining to fluid-conducting (xylem and phloem) tissues in plants
vascular cylinder (syn. stele)—the cylinder of vascular tissue in stems or roots
vascular wilt disease—a xylem disease that disrupts normal uptake of water and minerals, resulting in wilting and yellowing of foliage
vector—a living organism (e.g., insect, mite, bird, higher animal, nematode, parasitic plant, and human) able to carry and transmit a pathogen and disseminate disease; in genetic engineering, a vector or cloning

vehicle that is a self-replicating DNA molecule, such as a plasmid or virus, used to introduce a fragment of foreign DNA into a host cell
vegetative—referring to somatic or asexual parts of a plant, which are not involved in sexual reproduction
vegetative propagation—asexual reproduction; in plants, the use of cuttings, bulbs, tubers, and other vegetative plant parts to grow new plants
vein—small vascular bundle in a leaf
vein banding—symptom of virus disease in which regions along veins are darker green than tissue between veins
vein clearing—disappearance of green color in or around leaf veins
ventral—pertaining to the front or anterior of a structure
vermiform—worm shaped
verticillate—branching so that the branches are borne at the same level on the hypha or support and grow obliquely upward
vessel—water-conducting structure of xylem tissue with pit openings in end walls
viable (n. viability)—the state of being alive; able to germinate, as seeds, fungus spores, sclerotia, etc.; capable of growth
virion—complete virus particle
viroid—an infectious, nonencapsidated (naked) circular, single-stranded RNA
virulence—degree or measure of pathogenicity; relative capacity to cause disease
virulent—highly pathogenic; having the capacity to cause severe disease (*see* avirulent)
viruliferous—virus-laden, usually applied to insects or nematodes as vectors
virus—a submicroscopic, intracellular, obligate parasite consisting of a core of infectious nucleic acid (either RNA or DNA) usually surrounded by a protein coat
volunteer—self-set plant; plant seeded by chance

water-soaked—describing disease symptom of plants or lesions that appear wet, dark, and usually sunken and translucent
whorl—circular arrangement of like parts
wilt—drooping of leaves and stems from lack of water (inadequate water supply or excessive transpiration); vascular disease that interrupts normal water uptake
witch's-broom—disease symptom characterized by an abnormal, massed, brushlike development of many weak shoots arising at or close to the same point

xylem—water- and mineral-conducting, food-storing, supporting tissue of a plant

yellows—disease characterized by chlorosis and stunting of the host plant

zoosporangium—sporangium, or spore case, that bears zoospores
zoospore—fungal spore with flagella, capable of locomotion in water

Index